Colonel Augustus G. Hazard
and
The Hazard Powder Company
The Manufacture of Black Gunpowder

Edward Channing Allen, II

INFINITY PUBLISHING

ISBN 978-0-7414-7281-6 Paperback
ISBN 978-0-7414-7282-3 Hardcover
ISBN 978-0-7414-6750-6 Hardcover Color
ISBN 0-7414-9521-X eBook

Printed in the United States of America

Published March 2022

INFINITY PUBLISHING
1094 New DeHaven Street, Suite 100
West Conshohocken, PA 19428-2713
Toll-free (877) BUY BOOK
Local Phone (610) 941-9999
Fax (610) 941-9959
Info@buybooksontheweb.com
www.buybooksontheweb.com

To the men who worked in the Hazard Powder mills and made their homes in Hazardville.

TABLE OF CONTENTS

FOREWORD

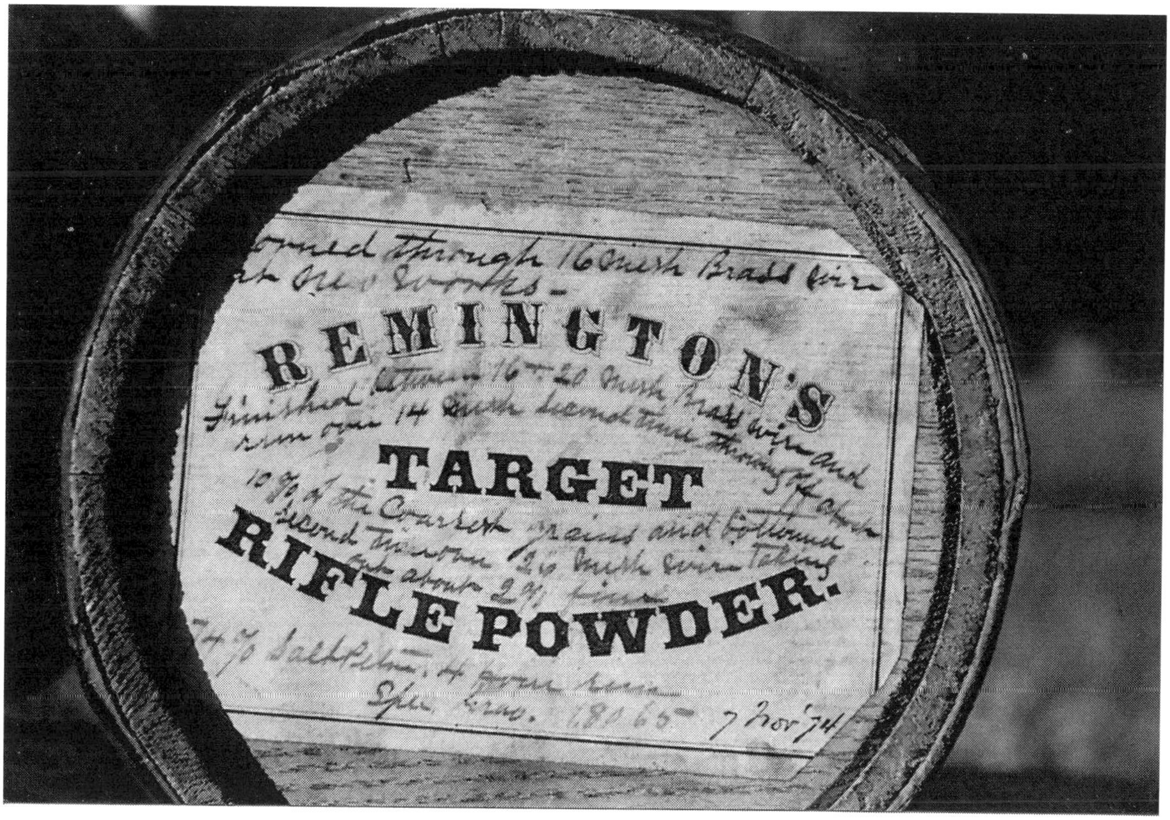

Foreword

Edward Channing Allen, II, was born in Stafford Springs, CT on December 22, 1927 to Herbert Walden Allen and Evalene Louiza Smith Allen. The Allen family lived on Main Street in the Hazardville section of Enfield, CT in the historic Olmstead house directly across the street from the Methodist Church. Edward's grandfather, Edward Channing Allen, I, purchased the 15-room house and settled there in 1888. The house was occupied by the Allen family for the next 112 years.

In 1882, Edward, I, bought the country store just east of the church and operated the business known as E. C. Allen's until his death in 1933. He provided many of the needs of the community during his lifetime from groceries to hardware goods and he was well respected by his friends and neighbors. After his death in 1933, his two sons, George and Herbert took over the operation and some years later George's son, Arthur, and Herbert's son, Edward, managed the business until closing in 1993.

As a boy and young adult, Edward, II, became interested in the history of the village and especially in the gunpowder industry that had been so prominent in Powder Hollow. He began studying Col. Augustus G. Hazard and the Hazard Powder Company, exploring the woods, rivers and trails in the hollow and pursued a lifelong interest in the black gunpowder industry. His college term paper, "Connecticut's Powder Hollow" was published in 1947, in the September and October issues of Connecticut Magazine.

Edward Channing Allen, II, Historian and Photographer at the Scantic River

His slide show "The Legend of Powder Hollow" consisted of 100 photographs, most of them copies of 19th century prints made in Powder Hollow when the industry was still active. He was invited to present his program to numerous historical and social groups throughout Connecticut and Massachusetts.

Edward was a former President of the Springfield Photographic Society, and won the 1953-1954 award for 'Color Slide of the Year'. He was a Star Exhibitor in the Photographic Society of America, past Chairman of the Springfield Color Slide Exhibit and former Vice-President and Master Member of the New England Camera Club Council.

In 1979 he presented his "Legend of Powder Hollow" slide show to the Fleur de Lis Museum in Faversham, England. Many of the powder makers that worked in the Hazard Powder Company came from Faversham. On our first visit to that town, we met and became long-time friends with the then President of their museum, a man named Arthur Allen (no relation to the Hazardville Allens) and his wife, Joan.

Edward and I were surprised to learn, on our first visit to the Hagley Museum in Wilmington, Del., that there were over 3,000 letters on file from Col. Hazard, and other Hazard Powder Company officers to the DuPonts. On that, and many subsequent visits, we spent many hours reading almost all of that correspondence.

Despite the hundreds of times Edward spoke on the history of the Hazard Powder Company, he was reluctant to put his lecture into print, because it changed with almost every presentation, adding new slides and updated information. However, he knew the historic importance of bringing the story of the Hazard Powder Company to life and felt the obligation, as the unofficial historian of Hazardville, Conn. to do just that. He started his compilation some years ago, and worked diligently on this book for months. He had the knowledge and facts surrounding the Powder Company to complete this task.

However, in 2005, Edward was diagnosed with a rare and incurable disease, and although his prognosis was dire, he worked as much as he could on his lifetime project. Edward died on March 29, 2007 without completing his work.

Edward Channing Allen, II, author
Compiled by Alice Godard Allen

In order to preserve all of Edward's research, pictures and personal knowledge, I pledged to finish the work he started and have put together all of the components that he had ready and present this book in honor of him. The true history of the Hazard Powder Company must be preserved for future generations.

In part one of the book is all of the text that Edward wrote, plus pictures, letters and other documents. Where necessary there are additions to complete the story. There are other pictures taken in Powder Hollow from 19th century glass negatives that heretofore, I suspect, have never been seen by the general public. Some were identified and some were not and they are included for the enjoyment of the reader.

Part two includes some little known vintage stories and pictures of Hazardville, Connecticut, which I have compiled in order to preserve this untold history. It is my hope that this book will serve as an historic document and find a place in the Hazardville Institute, the Enfield Public Library, the Enfield Historical Society, and in the homes of people who cherish the history of our forebears.

Alice Godard Allen

Introduction

Our Firm Foundations

Edward C. Allen, Enfield Historical Society

A Connecticut tobacco farmer might compare Enfield's foundations to the piers of a tobacco shed, each sturdy but interdependent. To the north was a manufacturing village named for Orrin Thompson. To the east, Wallop and Shakers were primarily agricultural. The central pier was comprised of small mills along the Scantic River. These mills were used for sawing, grinding, and carding, for distilling gin and producing gunpowder. The community of Enfield in the southern part of the town was the first area settled and the governmental and religious center.

Enfield or Freshwater Plantation, as it was first called, was the joint venture of five persons: Major John Pynchon, Samuel Marshfield, Thomas Stebbins, Jonathan Burt, and Benjamin Parsons. On October 31, 1579 these men were authorized by the Town of Springfield and the royal Government of Massachusetts Bay to make grants of land for homesteading "at the falls" of the Connecticut River. Settlers were required to establish themselves within three years or lose their grants. Two brothers, John and Robert Pease were the first to accept these conditions. Original settlers of Salem, they came to Freshwater Plantation, where they built a rudimentary hut near where the old meeting house formerly stood. They lived there through the winter alone. The next spring they moved their families and commenced the settlement, during the succeeding three years they were joined by nearly thirty families from Salem and vicinity.

Within the next year the Town Plat was laid out reserving land for roads and commons, for the ministry, for schools and other purposes. North Road (Hazard Ave.) was considered the north end of town. All of Somers was part of the tract. Each grant included a town lot, 30 to 60 acres of field land and a proportional share of the meadow. By 1684 regular religious meetings were being held, John Pease had been named first constable, and his daughter Margaret had become the first child born in the town.

This was also the year the General Court at Boston agreed to the petition for the Incorporation of the Plantation with Town privileges by the name of Enfield. Further title was obtained in 1680 when a "purchase was made of the Podunk Indian Sachem Notatuck of all the foot of the falls then extending eastwardly a full eight miles. The consideration was 25 Pounds Sterling."

In 1754 Enfield joined with Suffield and Woodstock in secession from Massachusetts and union with the Charter government of Connecticut. This transfer of allegiance was extraordinary in that it was without the consent of Massachusetts which had governed Enfield for seventy years. It was greatly approved by the residents however. Dr. John Pease stated in 1829, "...no intelligent inhabitant is insensible of the superiority of the institution of Connecticut over those of Massachusetts or has not often contrasted the excellence of our own cheap simple sad republican municipal regulations with the expensive and aristocratic systems of Massachusetts."

The Shaker movement came to Enfield in 1780. The "Shaking Quakers" lived an experiment in Christian communism. Each member of the sect turned over all his worldly goods upon joining the celibate United Society.

Enfield's Shaker settlement was grouped into North, South, East and West, and Church Families. By 1830 there were 260 Shakers on some 3000 acres in Enfield. One account said, "Their buildings are remarkably neat and convenient, and everything appears a model of neatness and economy. They are simple and plain in their manners, sober and industrious. There may be perhaps about 50 buildings in the settlement, consisting of dwelling houses, workshops, storehouses, etc. They also carry on various kinds of mechanical businesses, and their wares are such esteemed, being good and free from deception."

In addition to some of Enfield's most fertile farmland, the property included reservoirs or mill ponds now Shaker Pines Lake and Crescent Lake. Some craftsmen were assigned to furniture making, some to raising fruits, vegetables and herbs. Enfield became the home of the first packaged seed business in America as the best seeds of the Shaker gardens were bagged in cloth sacks for sale throughout the country. Much of the Shaker property is now occupied by the Connecticut State Prison.

The village of Hazardville takes its name from Col. Augustus G. Hazard, founder of the black gunpowder mills on the Scantic River. Hazard built an industry whose plant covered 400 acres in eastern Enfield and four miles of canals, 14 miles of roads and 118 powder mills. The mills took advantage of the river valley, using the year round supply of water power, and letting the hills isolate each stage of manufacturing for safety reasons. As the Hazard plant increased in size it expanded its market to the West Coast, and began to ship powder down the river for international use. The Crimean War saw both Great Britain and Russia, the antagonists, using the product of the Enfield mills.

As the Civil War approached, Hazard was producing 12 tons of sporting powder and 16 tons of blasting powder daily. There was some concern as to Col. Hazard's sympathies, for he had been an outspoken opponent of abolition and had entertained Secretary of War Jefferson Davis at the Hazard mansion on Enfield Street.

But when Fort Sumter was fired upon (with Hazard powder!) it was Hazard who addressed the town meeting at the Old Town Hall, pledging the support of his plant to the Union, and urging Enfield to send funds and men for the militia to help put down this insurrection. Ordnance records show that the Hazard Powder company supplied 40% of the munitions used during the war (by the Union forces). Gunpowder was produced in Hazardville until 1913 when an explosion so damaged the plant that it was closed by the Dupont interests who then controlled it.

In Thompsonville we find another example of a community being created around and named for the ingenuity of a man, Orrin Thompson. Thompson spent his childhood in Enfield, but like Hazard found himself after the War of 1812 in New York City "where the action was." He went to work for the firm of David Andrews and Company, carpet merchants. As he learned more about carpets he decided to try their production. In 1828 he built his first mill on Freshwater river. At this time the manufacturing village of Thompsonville consisted of 30 dwellings, two taverns, two stores, and five factories. Thompson's plant, "The White Mill." soon housed some of Brastus Bigelow's power looms. Bigelow's invention transformed the carpet business from a hand craft to a mass production operation.

Meanwhile, John Sanford was starting a similar venture in Amsterdam, New York. These two giants were to merge in 1929. From Orrin Thompson's 150 hands, Bigelow-Sanford at peak employed some 7500 persons, and Enfield was thought of as the home of the largest carpet plant in the world. While Hazard's home on Enfield Street became an inn, Thompson's Enfield Street mansion became the home of the Felician Sisters religious order. This was the foundation of Enfield, a residential community with a firm agricultural and industrial base.

Connecticut's Powder Hollow

THE HAZARD POWDER COMPANY

Connecticut's Powder Hollow – The Hazard Powder Company

by Edward Channing Allen, II

(This article appeared in Connecticut Circle magazine – Sept. & Oct. 1947)

To New Englanders, long familiar with the term "munitions plant", the fact that the Secretary of War commended a Connecticut explosives manufacturer for "ingenuity and courage with which the establishment was kept up despite occasional explosions" may not seem unusual. This was not, however, the occasion of a modern Army-Navy "E" award, but a citation made by Jefferson Davis, Secretary of War under Pierce's administration (1853-1857) to the Hazard Powder Company in Hazardville, Connecticut.

It was a Yankee Peddler, conceiving the idea of manufacturing the powder which was one of his chief articles in trade, who launched the Hazardville company. Abandoning his Vermont and New York selling circuit, Suffield-born Allen Loomis in 1835 bought the quiet meadows around the Scantic River in Enfield. Here in "Powder Hollow," as the spot was dubbed and is still known locally, the first set of mills operated by water power began grinding out the highly explosive dust at the rate of 20 twenty-five pound kegs a day.

Colonel Augustus G. Hazard, whom Loomis met in New York, became associated with the enterprise two years later and soon became controlling member. Colonel Hazard erected a mansion on Enfield Street, which was still the center of town, and settled down to take an active part in the business. In the spring of 1854 a banquet was held in the new hotel, and it was decided to call the new or western district of Enfield, Hazardville. In the summer of 1854 the Hazard Powder Company was incorporated by the Legislature with Augustus Hazard, Peter Cornell and Andrew Douglas among the incorporators. Loomis sold his interest in the corporation at about this time and Hazard was elected President.

The new company, under the direction of dynamic executive Hazard, immediately started construction of two stone dams, several ponds and many miles of canals for conveying the water power. The plant, covering 400 acres, included 200 separate buildings of which 118 were mills scattered over the lowland country. Improved machinery made use of the extensive water power available. By 1854 the first steam power works were running and the Horseshoe Dam was under construction. The production capacity of the plant at this time was 12 tons of sporting powder and 16 tons of blasting powder daily.

The Horseshoe Dam and South Maple Street bridge.

"The material for making 100 pounds of sporting powder," explained a life-long employee of the company, "was 75 pounds of refined saltpeter, 15 pounds brimstone (sulfur) and 10 pounds of charcoal, peeled alder cut on Somers Mountain baked in air-tight cylinders for about twelve hours. The charcoal, brimstone and several pounds of iron bullets around the size of large walnuts were put in iron cylinders placed in rows on a heavy shaft and revolved. In a few hours the charcoal and brimstone were ground as fine as flour." The mixture was carefully sifted to remove any foreign matter. At the same time the saltpeter was boiled with melted glue in vats five feet deep and ten feet in diameter. This refined product was cooled and washed in wooden vats before being mixed with the charcoal and brimstone.

"This was done," the powder plant worker continued, "in what were called wheel mills. The mixture was placed on a three-ton iron bed and mixed and ground by two seven-ton iron wheels continually turning in a circular fashion. In about four and one half-hours when the powder was 'ripe', it was taken from the wheel mill to the press where it was compressed into a cake about one inch thick and two feet square. "Then the powder was ready for the cracking mill where it was run through metal rollers covered with large teeth. This broke the powder into small pieces called grains. The powder then went to the glaze to be polished. About 5,000 pounds of powder with a small quantity of black lead were poured into large wooden cylinders. These tubs were set in motion for several hours for polishing. The grains were then screened for size and packed in shipping canisters and kegs ranging in size from ¼ pound to 125 pounds".

To plant superintendents went much of the credit for high production rates. Mathewson, the first superintendent of the powder mills, was followed by a Virginian named Garische, who believed that he could mix powder by a dry process. On Magazine Pond, one of the two reservoirs into which water from the canals emptied, a special mill was built for the experiment. Garische, a foreman and two others were killed in the explosion of this mill, and the fifteen room superintendent's mansion, built among the stumps of a clearing in what is now the center of town, was never occupied by the fair-haired boy Garische, for whom it was intended, nor by Colonel Hazard's daughter, whom it was rumored Garische was to marry. Instead, Edward Prickett, succeeding the unfortunate Garische, moved into the mansion and, with his son retained the managerial position as long as powder was produced in Hazardville.

Fairlawn, the Superintendent's Mansion

Fairlawn – Home of Supt. Edward Prickett
Second Mansion built for the Manager of the Company

Although every precaution was taken to prevent such accidents, the frequency of explosions remained the greatest hardship in the manufacture of black powder. The buildings were unheated and unlighted, smoking strictly prohibited, and the men required to wear shoes tapped with wooden pegs because of the danger of a spark struck by nails combusting the explosive black dust constantly in the air of the buildings.

As the mills were widely separated to prevent chain explosions, hand-car tracks of steel were used in transporting powder from one mill to the next. After lightening had struck one mill and carried along the tracks to others, the steel rails were scrapped. Stone and wood retaining walls augmented the natural hills in reducing explosion shock. Finally, the mills themselves were built in such a fashion that an explosion would cause the least possible damage. The framework of the buildings was constructed of heavy timbers rigidly connected, but the clapboards were loosely nailed in a vertical position so that they would fold down and out in the event of an explosion, leaving the framework intact.

Explosion and fire at a powder mill.

Despite all precautions 67 employees were killed in explosions of the thirty-five mills between March 11, 1839 and January 14, 1913. Wheel mills exploded 15 times, graining mills 14 times, press mills 6 times and mixing mills 3 times. Colonel Hazard's son Horace was killed testing a sample of powder at the mortar. One mill hand died when his powder-filled clothing caught fire.

The mortar at which Col. Hazard's son Horace was killed.

In the most disastrous explosion of the industry's history five mills exploded killing eight men and one woman, July 23, 1862. The first explosion at a press was followed in succession by explosions at another press, a graining mill and two wheel mills. The woman, running across an open field, was killed instantly by a stone hurled from the blast. On January 14, 1913 a ton and a half of powder blew up in four separate explosions killing two men, injuring others and wrecking four large mills. The roar was heard ten miles away.

"Chimneys toppled, glass was shattered, blazing fragments were hurled in all directions," runs a contemporary account. "Flying trees shot through the air. A traveler at a distance saw a wrecked roof sailing high in the sky. The explosion was felt with force in Springfield; housewives reported the rattling of dishes in their china closets."

After such an explosion, panic-stricken housewives gathered in knots on the street corners until casualty lists could be brought from the Hollow. An immediate censorship was thrown over the entire area until an investigation could be made. On one occasion an amateur photographer who rushed toward the rising plume of smoke where the "blow" had occurred was halted by Supt. Prickett who demanded his film. Such publicity was unwelcome.

Though deaths were many, jobs in the mills were prized. The work consisted mainly of loading charges in the mills, wetting them down occasionally and then leaving them to run themselves. Since the mills were unheated, the bulk of the long working day was spent in warm shacks where cards and relaxation ruled. Comparatively high wages lent attraction to the work for men with a zest for adventure. Before the burial of one explosion victim, the deceased's brother appeared at the superintendent's door applying for the job. The stories of miraculous escapes became legends in Powder Hollow. Several men who were given up for lost came home to their grieving families; they had been blown clear of the wreckage and in the panic following the explosion had been overlooked, so the story goes. According to another, a wagon driver had gone about a mile from the mill where he had taken on a load of kegged powder when a terrific explosion jarred him from his seat. Returning after delivering his dangerous cargo, he discovered a blast had destroyed the mill at which he had recently loaded.

All powder was carted directly to its destination or shipped by water from Enfield, prior to the opening of the railroad in Scitico. The rail line through Thompsonville would not accept gunpowder for shipment. From a dock below the Enfield Dam four flat-bottomed boats operated by the Douglas Company of Suffield hauled the explosives to Hartford. The life of these bargemen was an exciting one. Despite their fatalistic attitude – they often smoked while sitting on a shipment of powder – there was never an accident on one of the boats.

Big canvas covered wagons brought the powder from Hazardville, down the steep hill to the wharf where there was often enough explosive to ruin Enfield and all her neighbors. Powder was stored in a magazine near the wharf until the people vetoed that practice.

One sunny summer day the barge men and a group of loiterers were waiting on the wharf for one more load to fill the consignment of the two boats loading for a down river trip. "They saw the team come over the crest of the hill and begin its descent," a witness later related. "The load was so heavy the horses could not hold it back. The horses started to trot. A red keg of powder rolled off the front of the wagon and under the wheels. Some of the men fell over into the water to avoid the coming shock, for they felt sure there was to be a terrific explosion and that this would be about the last thing they would ever feel. Those who thought they were going to take a quick and somewhat hurried trip to Suffield and knew there was no use in dodging sat still and watched the wagon. The driver was powerless to hold the horses. The front wheels struck the keg, climbed on top of it. There was a little crash as the staves gave way. The black grains of powder scattered over the rocky road and the rear wheels passed over it, but there was no explosion. It was claimed by those who saw the sight that it took an hour and a half for the whole thing to happen, but the driver said that it took place in 11 seconds."

On another occasion barges were being unloaded at Hartford, a man standing on the deck taking the kegs from a man in the hold and in turn rolling them on to a third in a schooner laying along side. The schooner was nearly loaded and the men were hurrying to take advantage of a favorable breeze. As one of the kegs was rolled along the rail until the man on the schooner could take it, a hoop loosened and a thin line of powder grains fell in front of the rolling keg. There was a bright red flame. An explosion would destroy the ships and wreck the Hartford waterfront. But there was no explosion. It was just another queer quirk of the unpredictable black powder.

Powder Hollow took on new importance as Hazard, one of the "big three" in powder making for the Union, ran at full capacity night and day to meet increased demand during and after the Civil War. Producing competitively with DuPont in Delaware and Laflin & Rand in New York, the Hazard Powder Company reached the peak of its success.

Such business prosperity emphasized the blow to the industry when Colonel Hazard died at the Astor House in New York City, May 7, 1868. He left his fortune tied up in railroad stock and his mill at Hazardville in the hands of three men, two of them old and one ill. The invention of improved rifles had cut sales of the Hazard sporting powder which, with its excellent reputation, had been the most profitable manufacture. Blasting powder, a secondary product, was being replaced by the less dangerous dynamite.

The explosion of 1871 and the nationwide panic had jolted the company; the management was in distress. In 1876 General Henry DuPont, with whom the management had previously joined in forming the "Powder Trust," bought control of the Hazard Company in the largest powder deal negotiated in America up to that time.

The company maintained a nominal existence for some years, but few if any of the stockholders lived in Hazardville. A few years before powder-making ceased in Enfield, the Hazard Powder Company was dissolved and the business was absorbed by the DuPont Syndicate. When the Syndicate was forced by the government to dissolve, only local interest watched the Hazardville plant become a part of the Hercules Company.

The horse barn in powder hollow – once a riding academy

Today the building, which once stabled twenty draft horses, has become a riding academy. Hoof beats break the silence of the wooded river valley and young riders speculate on the origin of the overgrown, crumbling masonry of Powder Hollow.

Edward Channing Allen, III, circa 1960

COLONEL AUGUSTUS G. HAZARD
AND
THE EXPLOSIVES INDUSTRY

Colonel Augustus G. Hazard and the Explosives Industry

In 1850 the state of South Carolina formed a committee to oversee the protection of the state. In 1851, ten years before Fort Sumter was fired upon, Major James H. Trapier ordered guns, mortars, columbiads and howitzers for their arsenal. The order for gunpowder went to the Hazard Powder Company in Enfield, Connecticut.

The key to this sale of black gunpowder is tied to the adventures in the south of one Augustus George Hazard. Born in Kingston, Rhode Island in 1802, Augustus came from one of the founding families of Rhode Island. Augustus' father, Thomas, a sea captain, moved his family to Columbia, Connecticut where he took up house painting. Augustus, then six, soon learned about paints and oils while helping his father. At fifteen, Augustus left the farm, taking to the sea.

On August 1, 1821, the postmaster of Savannah published a "List of Letters Remaining in the Post Office". Near the bottom of the list was the name Augustus G. Hazard. Augustus Hazard, 19, had established himself in the South.

On December 2, 1826, the Savannah Georgian carried this notice:

FENCIBLES ATTENTION!

Appear on your parade ground in complete uniform,
on MONDAY, the 4th inst. at 3 o'clock P.M. for Company Drill
By order -A. G. Hazard -1st Serg't SF

Serg't Hazard can now be identified as the "Col. Hazard" who supplied this gunpowder from his mills on the Scantic River in Enfield, Connecticut, based on the Savannah-Charleston connection. The Savannah Fencibles (SF) were one of a number of militia units active in the Savannah area. A search of their records shows no one with higher rank than Captain, but Trumbull's HISTORY OF HARTFORD COUNTY refers to his Colonel's rank being bestowed by the Georgia military.

By February 20, 1827, he had established in Savannah a partnership with Allen Denslow to specialize in paints and sundries.

PAINT OIL, &c.

JUST received per ship Augusta. Brig

Pheasant and schr. Glide,
600 gallons Linseed Oil
200 do. Sets. Turpentine
2 tons Ground White Lead
30 kegs Verdigris green ground
20 canisters do do do
IN STORE
1000 gallons Sperm Oil
500 do whale do
20 boxes Sperm Candles
Window Glass of all sizes
As also Picture and Coach glass
Glass, Brass, and Japanese Lamps
And a general assortment of Paints

Oils, Brushes, Varnishes
And Painter's Articles

For sale by
HAZARD AND DENSLOW

That Hazard still plied the family trade is evidenced by an entry in the Hazard and Denslow ledger for painting:

1826, Dec. 8 Dr. Rev. Cutter Crawford, Payne

6 Days Work @2.00	12.00
Mr. Hazard time Graining	2.00

In 1827 Hazard left Denslow in Savannah to run that end of the business and relocated in New York, establishing a commission house. By 1835 he was acting as agent for the gunpowder mills of the Loomis Brothers, located on the Scantic River in Enfield, Connecticut. Dr. David Allen owned the land around the Scantic River. He was born in Melrose, in East Windsor, was educated at Dartmouth, and was a surgeon in the War of the Revolution. Dr. Allen died in 1834, and was buried in a private tomb by the road leading to the Hollow. In the next year, 1835, his farm and meadows were secured for the site of the greatest manufacturing enterprise that ever existed in eastern Enfield.

At this time Allen Loomis lived in the western part of Suffield. He was a trader in the days when trade was conducted mostly by barter. Each year he visited Vermont and Northern New York and exchanged gunpowder, whips and cigars for furs. Returning with the furs, he packed them on boats in the canal at Windsor Locks and shipped them to New York where the Astor family was prominent in the fur trade. The whips he obtained in Westfield, but he conceived the idea of increasing his profits by himself manufacturing cigars and gunpowder.

Denslow, still in Georgia, was commissioned1st Lieutenant of the Phoenix Riflemen (created when the Fencibles and the Georgia Volunteers merged in 1830).

The financial crash of 1837 caused Hazard to expand his interest in the black gunpowder field. The small cylinder mill built on the Scantic River in Enfield, Conn. by Denslow, together with Parkes, Allen and Neeland Loomis was ready for improved technology and expansion. William and Henry Prickett had arrived from Faversham, England with experience at the John Hall & Son gunpowder works. Soon they had three wheel mills going. William went back to England, returning shortly with nine more powder makers. These nine had agreed to work on a one year contract, then be free to take up farm sites in Ohio. With Augustus aboard, the firm became known as Loomis, Hazard and Company. In 1843 Hazard and Denslow bought out the other partners and organized a joint stock company, under the name of the Hazard Powder Company. The value of the partnership was transferred for a consideration of $66,425.

Colonel Hazard was the principal owner and president, Denslow continued as superintendent. Hazard subsequently moved to Enfield and lived there for the rest of his life.

Dr. David Allen's tomb on the road to Powder Hollow

Note: *When Bud Hardiman built the houses on the west side of South Maple Street, he demolished the remainder of the then empty tomb but saved the headstone, seen here above the door, and brought it to the home of the author, Ed Allen, on School St. Ed eventually donated the stone to the Hazardville Institute to be displayed on the property of the Hazardville Institute.*

KENTUCKY RIFLE GUNPOWDER
Manufactured at
Hazardville, Conn. by the
HAZARD POWDER CO.
Branch Offices:
Principal Office
NEW YORK.
HAZARD'S
GUNPOWDER
Branch Offices:
CHICAGO, ILLS.
E.S. Rice Agent
CINCINNATI, O.
R.S. Waddell Agt.
CHATTANOOGA, TENN.
F.J. Waddell Agt.
St. Louis, Mo.
W. McBlair Agt.
BALTIMORE, MD.
H.P. Collins Agt.
DENVER, COL.
S.C. Madden Agt.
NEW ORLEANS, LA.
BUFFALO, N.Y.
Clinton Bidwell Agt.
PITTSBURGH, PA.
D.W.C. Bidwell & Co.
KANSAS CITY, MO.
AGENCIES AND FRESH STOCK AT PROMINENT MARKETS THROUGHOUT THE COUNTRY.
When ordering from Merchants specify HAZARDS and for Cartridges always insist that they contain HAZARD Powder.
IT SHOULD COST NO MORE THAN OTHER BRANDS.

COL. AUGUSTUS G. HAZARD

The Hazard Family

Augustus Hazard was born in South Kingstown, Rhode Island, April 28, 1802, the seventh generation of the American sea going family which included Adm. Oliver Hazard Peary. When Thomas and his wife Silence moved to Columbia, Connecticut with their eight children in 1808, Thomas was listed as a house painter. A search of property transfers in Columbia and Lebanon indicates the extended Hazard family owned considerable farm land in the southwest part of Columbia. Augustus, the youngest Hazard child, left Columbia at age fifteen, soon to resurface in Savannah, Georgia. As we have seen, house painting gave way to entrepreneurship and the commission house partnership with Allen Denslow of New Haven. The business flourished and Hazard moved to New York City in 1827 to manage the northern loop of the enterprise

The remarkable series of advertisements in The Georgian newspaper show prosperous growth, usually promoting recently arrived goods and naming the ship which delivered them. On the same page an ad for the named ship would indicate its impending departure for New York.

Augustus was now a married man, having married Salome Merrill of West Hartford, Connecticut. By 1837 the family included daughters Elvira, (born May 15, 1824; married Governor Alexander Hamilton Bullock of Worcester, Massachusetts), Georgiana, (born July 29, 1827; married Joseph Sexton of New York), Fanny, (born October 1, 1830; married Ephraim Ward Bond of Springfield, Massachusetts), Emily, (born June 24, 1834; married Francis Elihu Dakin of Freeport, Illinois), and finally, a son, Horace Hamilton Hazard, (born January 1, 1832). Two additional daughters died young, and a last son, Samuel Douglass Hazard was born August 15, 1844.

The years as resident agent in New York ended with the great financial crash of 1837. Losing their packet ships, he and Denslow closed their Savannah warehouses as well. The site had served them well. Situated where the Hyatt Hotel is now located, it was adjacent to Factor's Walk, where cotton bales were traded. Offices were at street level, warehousing typically two floors below, on the river.

Hazard paid off his debt, emerged with excellent credit, and decided to ally himself with the gunpowder manufacturers he and Denslow had dealt with in Enfield. Legend tells us he built his Italianate mansion on Enfield Street with the profits from the purchase and resale of one ship load of India saltpeter. (Refined saltpeter, or potassium nitrate, comprises about 75% of black gunpowder when mixed with 15% charcoal, and 10% sulphur). Centerpiece of the mansion was a carved cherry spiral staircase imported from England at a reported cost of $20,000. He enjoyed entertaining prominent members of the Whig Party, of which he was chairman of the state central committee. Daniel Webster was a personal friend. Another weekend guest was President Pierce's Secretary of War, none other than Jefferson Davis.

Secretary Davis' visit to Connecticut is detailed in THE Hartford Daily Times of September 6, 1853. "Colonel Jefferson Davis arrived in this city by the morning train of cars from the North, September 5th, having spent the Sabbath at the country seat of the Hon. A. G. Hazard, at Hazardville (Enfield). He was honored by a salute of fifteen guns, fired from the state Arsenal, under the superintendence of Col. Webb. "He visited all of the public institutions, Colt's and Sharp's Armories, the charter oak, and other points of interest in our beautiful city".

"At twelve o'clock a public dinner was tendered to Col. Davis at the City Hotel....Col. Davis then rose and responded to the Mayor's remarks.....for some half or three-quarters of an hour....." He spoke of his own political position as a States Rights man, claiming that state sovereignty was essential to true republicanism. In joining the confederacy (of states) and subsequently entering into a more complete and lasting union, did Connecticut surrender her state sovereignty? Would she today give up her well established rights as a state? Every son of Connecticut would respond with a hearty "No!" and this makes him a states rights man at heart, whatever his professions may be....

"These machines sir, (turning to Col. Colt), which we have seen today, are wonderful to us of the South and West – really wonderful, sir, they are the surprising results of remarkable ingenuity. Your Repeating Pistols, made in Hartford, are relied upon on the borders of Texas to protect men and women of that state from Indian depredations. Northern manufactures are used by southern families. These are the sure bonds of union. The marks of prosperity and happiness are all over New England"...(he goes on to say he could live happily in beautiful New England).

Col. Hazard then followed with a few practical and happy remarks, and closed with the following: "Bring your cotton from Mississippi, and we will furnish the spindles; give us the material, and we will furnish the manual to spin out a fibre of the Union, as long as eternity".

One could conclude that these two men, one Yankee, one Rebel, were both expressing the political creed of their background. Concerning Augustus, it was later perceived as the confederate party line, and together with his well known years in Savannah left him suspect to many northerners. The record after Fort Sumter shows that after the "rebellion, he was a staunch supporter of the Union cause".

On April 26, 1861, this letter appeared in The Hartford Courant:

> "To Hon, Wm. A. Buckingham,
> Governor of the State of Connecticut.
>
> "Dear Sir – Every patriot and every man that loves his country no doubt deeply deplores, as I do, from the very depths of my heart, the present condition of our National affairs. These United States, recently so prosperous and so happy, this great and mighty nation that has commanded the respect, and won the admiration of the whole civilized world, is, I fear, upon the brink of destruction. We are entering into a war, one section of the country with the other – a civil war, the worst and most dreadful of all wars. Would to God it were otherwise, but so it is, the crisis is upon us and must be met, the government must be sustained, the Capitol defended, and Constitution and the Union preserved. Our gallant troops of Connecticut are responding to your summons and are mustering to march to the defense of all these our greatest and dearest National interests, and to aid you in fitting them out to take the field. I beg to hand you herewith a sight draft upon the Secretary of my Company for $500, having previously contributed that same amount at our office in the city of New York for the same object there, making $1000, in aid of the cause from my Company; and I now stand ready to give additional amounts to fit out the troops that are preparing to leave my own town, to join the army and am, with the highest respect and esteem,
>
> Your obedient servant, A. G. Hazard".

Augustus Hazard's business and life style mentor was Col. Sam Colt, whose Italianate mansion, grounds, and armor survive near Wethersfield Avenue, in Hartford. Hazard's own mansion was a smaller version of "Armsmere", Colt's home. Augustus described Colt's home: "The house is set like a duck in a pond, in broad and attractive surroundings. Built of brownstone and brick....with very thick walls, making it cool in summer and warm in winter. The main building is 42 x 125 feet. The rooms on the ground floor are thirteen feet high, and on the second floor twelve feet. There are eight open fire places for wood, and two large furnaces. There are ten bedrooms on the second floor. The house is situated on 50 acres of land. Butter, cream, eggs, chickens and the finest vegetables are produced, as well as a commercial tobacco crop. There are accommodations for seven horses."

On May 1, 1848, Hazard joined the Congregational Church located just down Enfield Street from his home. The church was growing, and a decision was made to construct a new edifice in 1848. Col. Hazard became chairman of the building committee, oversaw the moving of the 1776 church building across the street where it continued to serve as town hall. He gave the whale oil chandelier for the sanctuary, having imported it from France, together with two carved wooden columns in the pulpit area. He continued his membership until the unrelenting preaching of an abolitionist minister drove him away for a time.

Whale Oil Lamp – Enfield Congregational Church Sanctuary

Carved wooden columns – Enfield Congregational Church

The 500 acre plant acquired by Hazard and Denslow encompassed the valley of the Scantic River as it passed through Enfield. The Scantic rises in neighboring Somers at Bald Mountain. Like most watercourses, it constituted a ready source of power. Several mills were located just below Culver's Pond. About 100 years earlier a colonial grant had been made to the Davis family. A granite dam was constructed, using locally quarried stone and constructed by slave labor. The sluiceway powered a saw mill, a shingle mill, and a grist mill. Later dams in Somersville and Scitico powered other enterprises.

The first black gunpowder was produced in 1835 at a small cylinder mill located at the falls at South Maple Street. William and Henry Prickett arrived from England in the spring of 1836, and by fall had three wheel mills going. This production was in the area east of the Scantic River bridge on South Maple Street. It is here that the river drops about eight feet from its sandstone ledge. The Pricketts recommended it as a location for the first of three power dams.

The sandstone dam constructed here fed canals on both north and south sides of the stream. The northerly canal after powering the original mixing mill, culverts under Maple Street passing the red horse barn, proceeds behind the alder drying sheds and blacksmith's shop, then powers the machine shop before turning south to the corning mill.

First cylinder mill at the falls at South Maple Street, looking east.

Cylinder Mill at the falls at South Maple Street, looking west.

Alder Drying Sheds

From the machine shop, power is provided to the Dust house (now the Sign Factory) by a long overhead belt system. The drive belt is protected by a roofed covering, and otherwise resembles power lines, being located at a second story level.

The Machine Shop

Inside view of the Machine Shop
Jacob Stocker (center) – Henry Rosenberger (right)

Back at South Maple Street, the screened entrance to the southern canal is still visible as part of the remnants of the dam. (There are no records of the source of these sandstone blocks, but it is likely they came from the quarries in East Longmeadow. Records of St. Mary's Church, a gift of Col. Hazard to his Anglican powder makers, show East Longmeadow as the supplier for their sandstone trim).

The canal entered a tunnel constructed of dry laid stone. This tunnel proceeds west, then south under Powder Hill Road. George "Duffy" Ryan used to tell of walking through the five foot tunnel of low water to catch eels. The canal emerges from the tunnel to become the water supply for the entire mill system south of the Scantic. It is easily viewed as it parallels Powder Hill Road on one side and the reservoir known as Greenland Pond on the other. When the canal flows under the road, Powder Hill Road takes a dogleg to the east.

Two other landmarks are visible here. The first is a wagon road to the east. This road was cut at a gentle gradient in the direction of the Scitico Powder Switch where gunpowder could be loaded on rail cars at the corner of Abbe and Broad Brook Roads.

Scitico Powder Switch

Just past the dogleg to the east of the road was the Gunpowder Proving Grounds. Heavy, anchored bolts remain where a mortar was secured, pointing the direction of Powder Hill. A standard test of cannon powder strength involved putting one ounce of powder in the mortar, followed by a 32 pound cannon ball. U.S. Ordnance required the ball must travel 300 yards to meet their standards.

Blasting Powder Dry House – Henry Rosenberger on Porch

As the canal continues south it passes another dust house, several mills, a dry house and the Steam Glaze. The canal ultimately becomes the supply for Magazine Pond, the largest of the man made storage reservoirs. The system proceeds back toward the Scantic, powering other mills, filling Greenland Pond and exiting to the river just beyond the Scantic Bridge.

Magazine Pond

Greenland Pond

The complex of mills to the east was supplied by a second sandstone dam constructed in 1842. This upper dam was located off Cooper Street at the "New Works." Again, water powered canals were constructed to the north and south of the dam. The northern canal cut into the steep valley wall. It first powered a glazing mill located near the end of today's Gordon Lane. Nearby were the cooper shops where wooden kegs were manufactured.

The Upper Falls

These shops prompted the lane to be called Cooper Street. Crossing under Cooper Street, the canal supplied the Upper double wheel mill and terminated at the fabled wheel mill always known as the "Man-Killer."

Upper Double Wheel Mill – a/k/a "The Man Killer"
Henry Rosenberger at wagon/ Tom Dowd on right

Contributing to this canal was a brook winding south from Hazardville. This brook originated at a spring fed pond in the Allen Street area. It ran as an open brook west of a machine shop where metal powder kegs were manufactured (later the location of Connecticut Valley Oil), then under the footings of the Hazardville Institute.

Col. Hazard's insistence on maintaining this source of water power is identifiable as the brook enters an arched tunnel in the Institute crawl space and into a tile beneath Hazard Avenue. Passing behind the Methodist Church and E. C. Allen's store, the brook re-emerges to join the canal system.

The canal south of the Cooper Street dam supplied the "New Works," a steam plant constructed after the Civil War. It was recognized as the most modern black gunpowder plant in the industry. Again, this canal supplied additional support to the mills before returning to the Scantic.

The New Works

While the Pricketts and most of the early powder makers were of English stock, construction of the canal system was said to have been accomplished by Irish immigrants who came to America during the great Irish potato famine. A third horseshoe shaped dam built immediately east of the bridge in 1854 served to recover the discharge from the Greenland Pond system as it returned to the river.

George Bridge came from Milton, England with his father and mother in 1842. A nephew of the Prickett boys, he wrote a paper describing the mills in 1843. "The original presses were worked by windlasses, then by hand pumps, and later by water power....in 1843 there were six wheel mills, two corning houses, two presses, a glaze house and a combined drying or 'hot' house and dusting house. A new drying house was built in 1844 and the old one was used for dusting alone for the next seventy years. In 1850 the first steam driven mills (or engine mills) were installed; and in 1850-1854 other mills were erected at the lower falls…new magazines and a saltpeter refinery were also built".

Water Canal

Miscellaneous Labels

MANUFACTURING BLACK GUNPOWDER

Manufacturing Black Gunpowder

Black gunpowder is created by physically combining three ingredients: saltpeter, sulfur and charcoal. Saltpeter (potassium nitrate) was shipped from Calcutta, sulfur from Sicily and the charcoal was produced locally by the controlled burning in the charcoal house of alder wood. Alder saplings became the spring "cash crop" for farmers in Wilbraham, Stafford and Somers. The two long buildings on Dust House Road were alder drying sheds, located across the street from the kilns. When the pollution from the burning kilns drifted to the north, it bestowed the name "Smoke Street" to what today is more properly called Southview Street.

Smoke from burning kilns drifting over Southview Street, a/k/a Smoke Street

The first step in the production process was the refining of the saltpeter and sulfur. This procedure was crucial, for the quality of powder depended upon the degree of refinement of the ingredients. Here the crude sulfur was heated in closed iron vessels called retorts. Passing from a liquid into a gas, it recondensed as a purified liquid that solidified as it cooled. Saltpeter was also refined here. It was boiled in a large kettle with water and a small amount of glue. Some impurities floated to the top and were skimmed off, while the heavier, unwanted salts fell to the bottom as residue. The purified, liquid saltpeter was then decanted and cooled in vats. Now white and solidified, the pure saltpeter was washed, dried and ready for pulverizing.

Saltpetre Refinery

Inside the refinery

Inside the refinery

Following the refining process, the sulfur and the charcoal were blended in the pulverizing mill to form a mixture called 'dust.' Because black powder is a mixture, not a compound, a thorough mixing is essential. This was accomplished in Powder Hollow in roll mills, which employed two, eight-ton iron mill wheels to produce a homogeneous mix. The water wheels were eventually replaced by more efficient water turbines.

View of the eight-ton iron wheels used in the mixing of gunpowder.

Crucial to the incorporating process was wetting down the chemicals – as much as 400 pounds – as they were spread on the bed plate. This process, called liquoring, depended upon the skill of the first mill hand to judge even the degree of humidity present. These mill runs were up to four hours long, and required periodic inspections. In the meantime, mill hands were permitted to retire to a warming hut.

Press House

After incorporation, the powder 'wheel cake' was taken to the press house where it was pressed hydraulically to increase its density. These cakes, two inches thick and two feet square, were transported on narrow gauge hand cars, sometimes called angel buggies, to the next mill site in the chain, perhaps one-quarter mile distant. Safety was always the first concern, so the iron rails ended near the mill, to be replaced by a section of oaken tracks.

Rufus Stratton and pressed cakes of gunpowder on an "Angel Buggy."
What looks like a piece of slate propped against the buggy is actually a cake of gunpowder.

Written on the cake is:

Kentucky Rifle
Powder
Press Cake
for
F, FF, FFF Grain

The press cake was next broken into 'chunks' in the chip-roll machine and carted to the corning or graining mills. Here the powder chunks were reduced to grains that were sorted by size. Coarser varieties would be marketed for cannon and blasting; the finer powders went for small arms.

Cracking or Corning Mill – Mr. Vasseur is pictured on wagon

In the next stage of processing, the powder was tumbled in glazing barrels to round the grains. Graphite was added to the barrels to make the powder shiny. The powder was then sent to the heated dry house to remove moisture. In the pack house, powder was given a final screening to remove any traces of dust created in the earlier processing.

Glaze Mill – Charles Vasseur on wagon and Matt Cunningham

Packing was done in the canister house approached on a carriage road lined with American Elm trees. Gunpowder was packed in canisters of one-half, one and five lb. sizes, and kegs of 6 ¼, 12 ½, and 25 lbs. The tin canisters were manufactured by John Law and Co. John Law had learned tin smithing from the Olmsted family.

Canisters inside the storage magazine.

The canisters were made on the second floor of the tin shop in a building just east of the Hazardville Institute building. The Laws also made iron kegs in the long building that stood south of Firtion's Garage (Ct Valley Oil). It is now located back of the E. C. Allen house. Prior to that all kegs were wooden ones.

The Packing House or Canister House

The canister house served as the shipping building. It stood where the pitcher's mound is located today in Powder Hollow Bowl.

Modern baseball field – former site of the Canister House

Shipping was long a problem for Augustus Hazard. The New Haven Rail Road, which ran between New Haven and Springfield, would never carry gunpowder. Consequently, Hazard powder had to be shipped by covered wagon or water for over forty years. In season, powder would be loaded on barges at the dock on Bridge Lane at the Connecticut River. These barges were then poled down river to Hartford where the powder kegs could be reloaded on to schooners. Hazard had storage magazines on Bridge Lane and in major sites across North America.

A Magazine Storage

There is much lore about the bravery of these barge men. In the event of a thunderstorm, it was customary that the barge would be anchored in the middle of the river, so that the barge men could take refuge on the shore. On one occasion, one pole man told his friends he was going below deck to wait out the storm. "The lightening can get you as soon on the canal bank as here on the barge." To be sure, lightening did strike the barge. There was no fire, no explosion, just killed the barge man.

A Hazard family feud surfaced in connection with the magazines located on Bridge Lane. Augustus Hazard's brother, Stanton, had moved to Enfield. Col. Hazard took an interest in Stanton's son George. It appears that the colonel helped provide George with an education, and he may have worked for the powder company. At any rate, there was a falling out. George attacked Col. Hazard by publishing a broadside entitled:

SAVE YOUR LIVES!
AND PROPERTY
CITIZENS OF ENFIELD AND SUFFIELD:

I deem it my duty to warn you are all liable at a moment's notice to be blown up into the air, yourselves, your families and your houses, through the illegal storage of immense quantities of Gunpowder at the Hazard Powder Company's Magazines at the River and at the Mills in Scitico. The careless manner in which these Magazines are managed is well known, and it is a matter of astonishment that no "accident" has yet happened. I have picked up matches on the floor of the river Magazine when to my knowledge there have been Ten Thousand Kegs of Gunpowder therein. The most appalling accidents by Steam would sink into insignificance compared with an accident at the magazine.

Read the following account of the consequences of Twenty Kegs of Gunpowder exploding, and then picture to yourself the scene of horrors if Ten thousand kegs exploded.

EXPLOSION OF POWDER

Baltimore, August 31, 1853

"A terrible explosion occurred at Wheeling on Monday. Twenty kegs of powder exploded on the wharf, killing one drayman instantly, and blowing another into the river. The wharf boat was completely demolished, and the steamboats Orion and Salem damaged. Windows were shattered through the city".

"Pause! Reflect! Fellow Citizens, on what would have resulted from one inadvertently stepping upon one of those matches – my blood even now curdles with horror at the bare thought. I can give you a very slight, imperfect idea of the consequence – every house in Enfield would be leveled with the ground; no breathing thing, man, bird or beast, within ten miles, but would be instantly killed in the concussion; Hartford and Springfield would be shaken as with an earthquake, and probably Fifty Millions of Dollars would not cover the damages – and all this frightful wholesale murder and destruction of Property would result from the Hazard Powder Company defying the Laws. The Law says they shall not store more then Two Hundred Kegs of Powder in one Building, yet they, in defiance of Law, put Fifteen Thousand Kegs in their Magazine at one time; and so reckless are they of consequences that they do not use sufficient precaution to prevent Lucifer Matches from laying about the floor.

"Fellow Citizens, most of you are aware that Augustus G. Hazard, Colonel of the Brighton Artillery, had injured me, and I am determined that he shall no longer defy the Laws of Connecticut. I do not deny that in consequence of his ill-treatment of me, I have personal feelings in compelling him to obey the Laws; But whatever my motives may be, the Public will receive the benefit of my exertions, and possibly I may be the means of saving thousands of Lives and Millions of Property from being destroyed by the illegal storage of large quantities of Powder in those nuisances, the Magazines of the Hazard Powder Co.

"I confidently trust that the Citizens of Enfield, Suffield and the neighboring Towns and Cities, more especially Hartford and Springfield, will give me their aid in suppressing this Nuisance, and in compelling The Hazard Powder Company to respect the Laws, the more especially as their lives and their property are at stake.

Enfield, September 13, 1853 GEORGE HAZARD

(Note: It should be noted that other publications erroneously identify George as Augustus G. Hazard's ***son***. George was actually A. G. Hazard's ***nephew***, son of A. G. H's. brother, Stanton.)

In respect to George's call to arms little is recorded. Perhaps Uncle Augustus was referring to George in a letter written to his superintendent Ed Prickett in January of 1851. In speaking of his own efforts to "lay the foundations of a village that I hope might some day grow up to a respectable size….I also desire to do what I could to promote the general interest, the prosperity, happiness & advancement of the whole Town". He goes on, "After spending more than twenty thousand dollars in trying to educate delicate (sic) poor relations, without I may say any success except in a single instance – I thought I would turn and direct my efforts of this kind (outside of my own family) to strangers or to others not related, and to those who would be more likely to appreciate well intentioned efforts".

Col. Hazard's letter, once termed his manifesto, rants on for nine pages, much of it dealing with his conservative white views and his distaste for the abolitionist movement. Written three months before the attack on Fort Sumter, the letter mentioned his success with "strangers and others," such as A. E. Douglass and William S. Colvin, both of whom became officers of the company.

SAVE YOUR LIVES!

AND PROPERTY.

CITIZENS OF ENFIELD AND SUFFIELD:—

I deem it my duty to warn you that you are all liable at a moment's notice to be blown up into the air, yourselves, your families and your houses, through the illegal storage of immense quantities of Gunpowder at the Hazard Powder Company's Magazines at the River and at the Mills in Scitico. The careless manner in which these Magazines are managed is well known, and it is a matter of astonishment that no "accident" has yet happened. I have picked up Matches on the floor of the River Magazine when to my knowledge there have been Ten Thousand Kegs of Gunpowder therein. The most appalling accidents by Steam would sink into insignificance compared with an accident at this Magazine.—Read the following account of the consequences of Twenty Kegs of Gunpowder exploding, and then picture to yourself the scene of horrors if Ten Thousand Kegs exploded.

EXPLOSION OF POWDER.

BALTIMORE, Aug. 31, 1853.

"A terrible explosion occurred at Wheeling on Monday. Twenty kegs of powder exploded on the wharf, killing one drayman instantly, and blowing another into the river. The wharf boat was completely demolished, and the steamboats Orion and Salem damaged. Windows were shattered throughout the city."

Pause! Reflect! fellow Citizens, on what would have resulted from some one inadvertently stepping upon one of these matches—my blood even now curdles with horror at the bare thought. I can give you a very slight, imperfect idea of the consequence—every house in Enfield would be levelled with the ground; no breathing thing, man, bird or beast, within ten miles, but would be instantly killed by the concussion; Hartford and Springfield would be shaken as with an earthquake, and probably Fifty Millions of Dollars would not cover the damages—and all this frightful wholesale murder and destruction of Property would result from the Hazard Powder Co. defying the Laws. The Law says they shall not store more than Two Hundred Kegs of Powder in one Building, yet they, in defiance of Law, put Fifteen Thousand Kegs in their Magazine at one time; and so reckless are they of consequences that they do not use sufficient precaution to prevent Lucifer Matches from laying about the floor.

Fellow Citizens, most of you are aware that Augustus G. Hazard, Colonel of the Brighton Artillery, has injured me, and I am determined that he shall no longer defy the Laws of Connecticut. I do not deny that in consequence of this ill-treatment of me, I have a personal feeling in compelling him to obey the Laws; but whatever my motives may be, the Public will receive the benefit of my exertions, and possibly I may be the means of saving thousands of Lives and Millions of Property from being destroyed by the illegal storage of Large quantities of Powder in those nuisances, the Magazines of the Hazard Powder Co.

I confidently trust that the Citizens of Enfield, Suffield, and the neighboring Towns and Cities, more especially Hartford and Springfield, will give me their aid and support in suppressing this Nuisance, and in compelling the Hazard Powder Company to respect the Laws, the more especially as their lives and their property are at stake.

GEORGE HAZARD.

Enfield, Sept. 13, 1853.

Toward the end of his life, William Colvin described how Col. Hazard changed his life:

Sunday, April 9, 1922

Dear Brother Henry,

You have taken such an interest in the account of my business career that I want to tell you of something which occurred on the same day that Col. Hazard arranged to send me to the New York Office of the Hazard Powder Company in October, 1860.

Joseph Sexton and Edward Prickett had gone to dinner. I was alone in the office and picked up a copy of the New York Illustrated News which someone had left. It contained several pictures of the best buildings, parks, and waterfront of New York, over which I got so excited that I went down on my knees right then and there and promised that before I came of age I would take William S. Colvin down to the big city to see all that was pictured and more.

Just as I finished I heard the rumble of wheels and saw Colonel Hazard in his carriage coming past the machine shop corner. On entering the office and seeing me he said, "William, how would you like to go to New York City?" My answer was prompt. "Col. Hazard, I would love to see the big city." "Oh," he said, "I mean to go to live, to grow up in our New York Office?" I said I would have to talk it over with Mother when I went to dinner. Well, to make a long story short, I got Mother's consent. Col. Hazard ordered Mr. Sexton to send me to Springfield and gave me enough money to have Tilly Haines and Company fit me out with a suit of clothes, city style, and the following week I was on my way to New York. Douglas Hazard, the Colonel's son, accompanied me as far as New Haven, where he was to enter Yale College and where he was stricken with typhoid fever and died shortly after."

The Colvin family of powder makers were unusual among the English immigrants in that they trained at the Hounslow Mills near London rather than at John Hall and Son near Faversham, Kent. William Samuel Colvin was born at Twickenham, England, on December 16, 1841, and died at Brooklyn, New York on October 6, 1922. He started at work at the Hazard mills in 1858 and became clerk and traveling agent in the New York office of the company in 1860. He was made treasurer in 1878, vice-president and treasurer in 1886, and president in 1892. He held this last position until his retirement in 1902. In addition he was a director in more than a dozen other powder companies in which the Hazard Powder Company had a stock interest.

William's gracious Victorian home was in Brooklyn Heights. One can readily picture him commuting across the Brooklyn Bridge to the home office at 89 Wall Street. Correspondence in the DuPont archives allude to trips to Europe, a perk never dreamed of by the youngster yearning to visit New York City. William's brother, Arthur E. Colvin, was also employed at the mills and lived out his long life as a Pensioner at Hazardville. Henry, the third brother, was employed at the U. S. Arsenal at Springfield, Massachusetts. Arthur G. Allen, a grandson of Arthur Colvin, survives as the only Colvin kin of his generation still residing in Hazardville.

MAP OF HAZARDVILLE AND THE HAZARD POWDER COMPANY

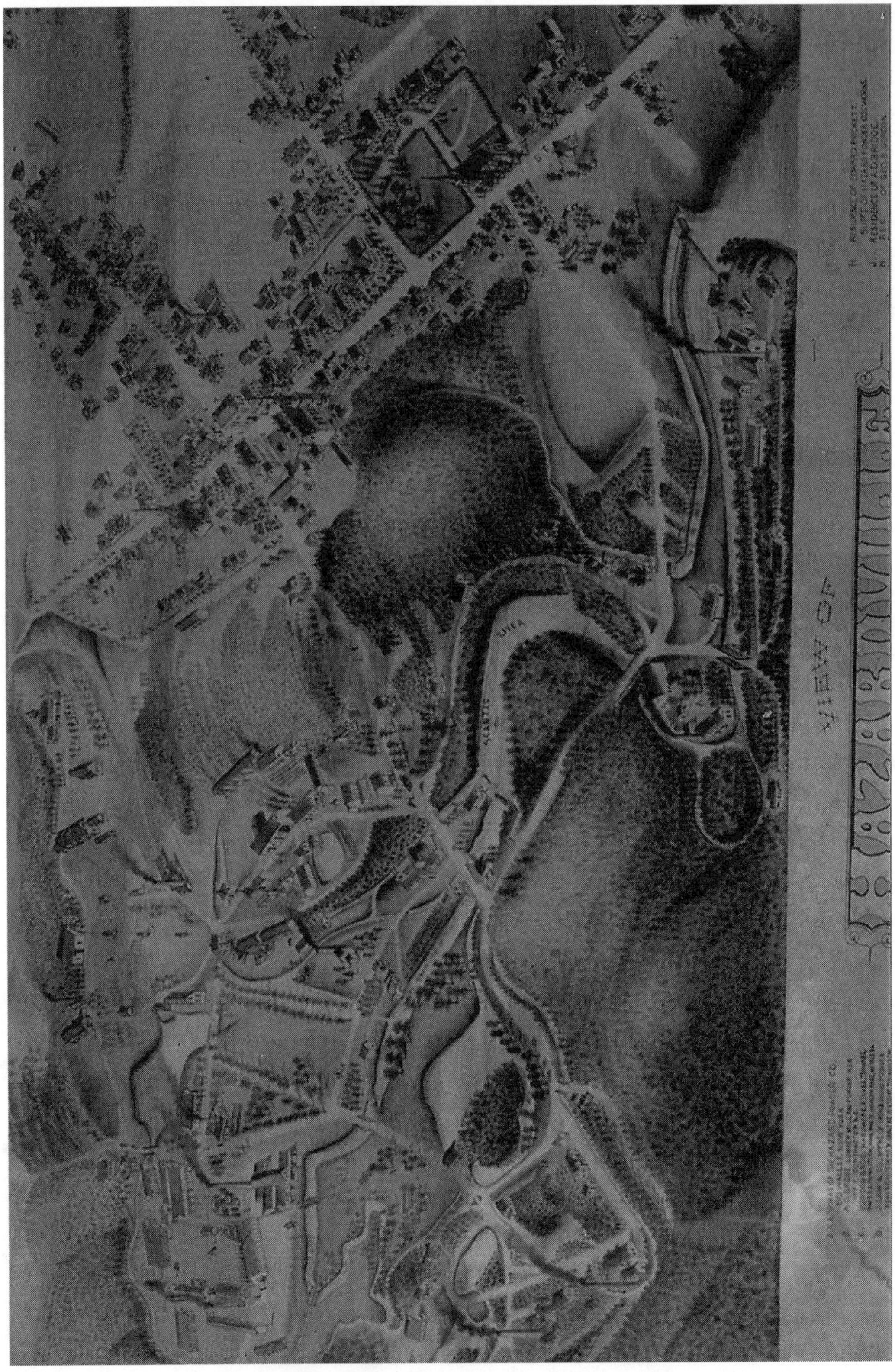

There were two Sporting Powder Lines, one starting at No. 1 and following down to Press 2, Corning Mill 3 and Glaze #4. The other consisted of double wheel mills 6 & 7 and single wheel mill #15 all run by water power. The press was down out of sight to the left. Corning Mill was #13 and then the powder went back and mixed at Glaze #4, next down to Sporting Dryhouse #29 then back to Glaze #44, then to Sorter #30 and Packhouses #33, 34 and 35. The Blasting Plant started at #23, Press #16, Corning Mill CM (left edge) Glaze #20, Dryhouse #18, Sorter #17, Pack Houses and Magazines not shown.

Some buildings are not marked as they were either vacant or stored obsolete material. The chimney and building near #44 was originally an Engine House but when this map was made there was only the chimney and shell of the building left and no smoke coming out.

The Scantic River had a dam at Somersville 2 miles east for a cloth mill, then Scitico Powder Plant, then two more dams to Scitico then 3 dams at the H. P. Co. plant in Hazardville and on down to Conn. River.

Compiled by E. L. Prickett, Jan. 24th, 1948

The Faversham Powder Makers

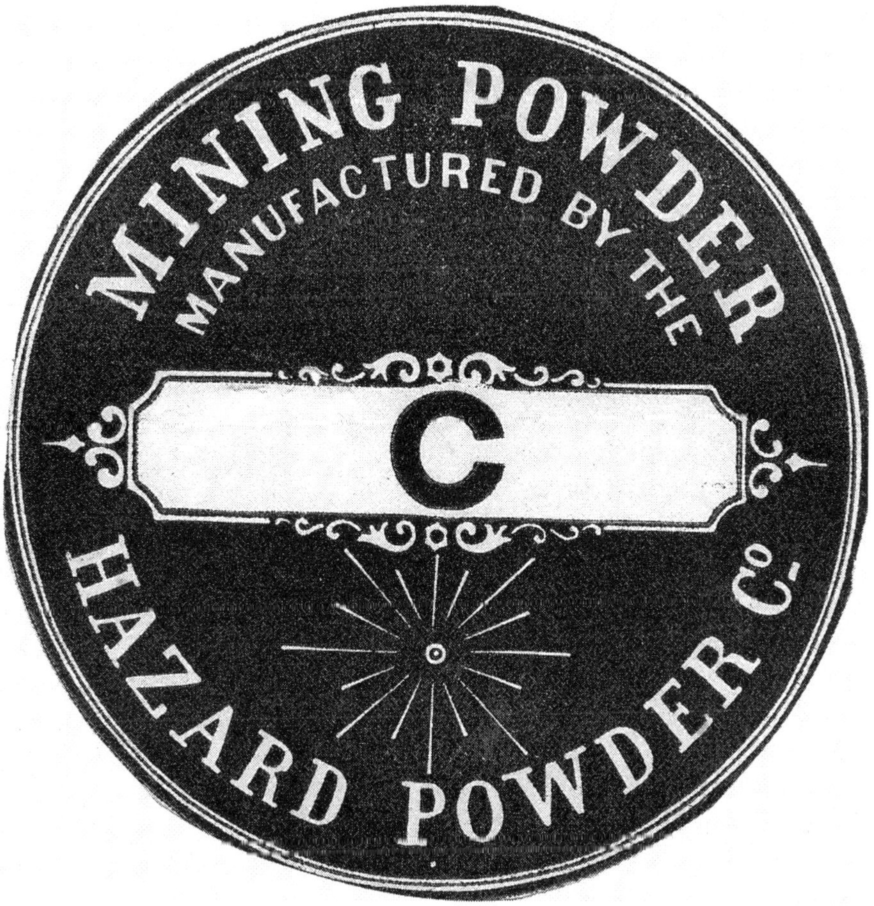

The first and foremost family of British powder makers to come to Connecticut were the Pricketts. "It is said that Thomas Prickett, (then named Thomas Fluellen, a Welch name) was a retainer of the King of England. They were on a hunting trip for a deer and the King said, "I will ennoble the first man to kill a deer". Thomas Fluellen was the first one of the party to do so, and in the ways of chivalry, the Monarch smote him on the back with the flat of his sword and proclaimed him "Sir Thomas Prickett". A young deer just sprouting his antlers is called a pricket. (Note The Bridge Family History, Marjorie Nadeau).

The arrival of Henry and William Prickett is described in the Prickett family correspondence that follows. Two other brothers, George and Jesse, were among those recruited shortly thereafter.

> "Of these four brothers, some of whom had worked in powder mills in England previous to their coming to this country in 1836, Henry, the oldest, born in Faversham in 1801, was severely burned in 1838 and later moved to Schaghticoke Powder Mills in New York state. In 1849 he moved to a farm in Illinois. William was a powder foreman and was killed in a wheel mill explosion in October, 1844 at Hazardville, Connecticut.
>
> "George (baptized 30, Sep., 1813 in Ospringe, Kent) was killed in a wheel mill explosion in 1842 at Hazardville. Jesse, the youngest (baptized 13, March, 1819, in Ospringe), left and moved to Saugerties, NY where he was killed in a corning mill of the Laflin Mills in 1840.
>
> "The sons of only one of these brothers continued to work in the powder mills at Hazardville and they were sons of William Prickett, they being Edward, Thomas, Ashmun, and William and each of these brothers had one or more sons who worked in the mills later on. Edward Prickett, also born at Ospringe, 22, Feb., 1832, was working in a canister house at the age of twelve in 1844 when his father was killed. He served as a cooper making powder barrels and kegs before he was employed permanently by the powder company in 1854 at the age of twenty-two, when he was taken into the office. "In 1859 he, with Jos. Sexton (son-in-law of Col. Hazard) were made joint superintendents and in 1867 he was made sole superintendent and served in that capacity until August, 1905, when he was pensioned at the age of 73.
>
> "Edward's eldest son, Clifford Denslow Prickett, whose middle name was in memory of the original founder, A. A. Denslow, entered the employ of the Hazard Powder Company in April, 1880 as a clerk, and in 1890 was made Assistant Superintendent. In 1904 he was transferred to the Wilmington office of the DuPont company as Assistant to Lammot DuPont. In 1906 he became assistant to George Patterson, General Superintendent of Black Powder Operations, and in 1912 he succeeded him at the time of his death. In 1913 when the Hercules Powder Company was newly formed, he was made one of the vice-presidents and directors, and for a number of years had charge of all Black Powder Plants. He still worked in the industry in 1936.
>
> "The second son of Edward Prickett, Edward L. Prickett, entered the employ of the Hazard Powder Company as a laborer April 1, 1888 and on May 1st was taken into the office as a clerk. On the retirement of Edward Prickett Sr. in August, 1905, Edward L. Prickett was made Superintendent at Hazardville, was transferred to Schaghticoke Plant in 1908, then transferred back to Hazardville in 1909 for six years. In 1915 it was back to Schaghticoke where he held that position until the closing of the plant, an actual service of 35 years.
>
> "Coming down to the fourth generation to be with the company at the end of 100 years.......there is only one Prickett and that is the son of Edward L. Prickett, Louis E. Prickett. Louis packed canisters and worked with the yard gang.....when a youngster, in 1925 began as a clerk in the Hercules Office. In 1936 he was still employed in the business, working in the Experimental Station in Wilmington, DE.
>
> "George Bridge, another Milton native, was the son of John Bridge and Mary Prickett Bridge. The Bridge family came to America with a promise of work for John "Canistering powder and pegging kegs" for $22 to $25 per month. This work was

"described as 'easy and but little danger the work is such as woman youse (sic) to do at Halls Mills, you had better apply quick.

W H G & J Prickett".

John and Mary Bridge sailed in March of 1842 on the ship "Mediator" with their five children. George, the eldest of their children, wrote the following account of their departure.

"We had all of our goods on the team on the night March 14, 1842. We started from Milton about 2 o'clock in the morning got there in the morning of 15th of March for London and got there in the morning of 17th of March. I remember seeing St. Paul's Cathedral on the way to the dock. We arrived in New York on April 25th, but did not land until the next day, and then not in time to start for Hartford on the steamboat. We stayed all night with an English family and I remember having some plum pudding. How good it tasted after living on Hardtack for six weeks. We started on April 27th, 1842 and arrived in Hartford early in the morning of April 28th and came by team to Scitico.

"Father got a team and a man to drive. The goods were all loaded on then we all got on, Father, Mother, George, John, Ruth, Amos and Ephram. The two boys had to walk up some of the hills. When we got to Enfield we saw a church in the middle of the road and just west of it a hotel. They told us it was three miles to this place. We came and we got there at 2 o'clock. First we came to a church in the woods. Then some men were at work raising the old tin shop, the same building is Andrew Gordon's barn. Then we came to the old brick house which stood between E. C. Allen's store and the M. E. Church. Then we met Aunt Betsy Prickett and she went back with us to her house which is down on Cooper Street or nearly opposite the Episcopal Church. They backed up the team and unloaded all our goods. We had two rooms and a garret, Uncle George Prickett and his family had two rooms and a hall bedroom.

"We lived on the corner of Cooper Street until the late summer of that same year, 1842, when we moved into the south part of the old Boarding House in the Hollow. We were living there on August 2, 1842, when Uncle George Prickett was badly burned in an explosion about ten o'clock in the morning and died about ten o'clock that same night. He died in the same bed John and I slept in. Another explosion occurred in June 1844 after a hard thunderstorm with hail.

"The explosion blew the doors off the old boarding house and broke nearly all of the windows. My mother was in the house at the time. She had just sent Ruth off to school and the child was only about as far as the tomb. One piece of timber came down close to Amos and lots of small pieces all about.

"Mother had a baby about eight months old in a cradle. It blew a door across and she thought it had killed Eben but the tram of the cradle protected him and he was not hurt much. After that my folks got out of the Hollow as soon as they could. In about a month they had an opportunity to move up into the east part of the old brick house on Main Street."

From a History of Powder Hollow by George Bridge, as quoted in "The Bridge Family History," 1983.

Government Test Box for Gauging Gunpowder

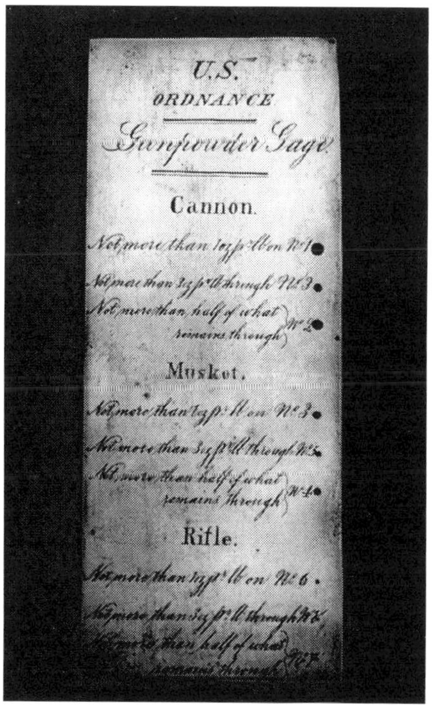

U. S. Ordnance Gunpowder Gage

View of the Packing House looking north.

Horseshoe and Upper Falls

MINING POWDER
MANUFACTURED BY THE
C
HAZARD POWDER CO.
BRIGHT GLAZED

The Civil War and Gunpowder

Chapter Three

A THESIS BY DAVID N. JELLISON

Abraham Lincoln had no sooner become President when, in the spring of 1861, the South decided to take possession of the one fort in Charleston Harbor still under federal control. This event took place on April twelfth of that year. Just ten days later Hazard had this to say to DuPont concerning the powder situation in the south while also informing him of an order he had already received from the United States Government.

> We have instructed all our agents in the South to sell what powder they have remaining at their regular prices but for cash only, since we cannot remove or control it.... Since writing foregoing letter have application for 1000 barrels of cannon powder and 500 barrels of musket powder. May be no time to confer with you shall prepare 18 cents per lb. (1)

In 1861 Hazard and DuPont consumed one-half the saltpeter used in America which indicates the magnitude of their powder making capacity at the beginning of the war. (2) The following description of the Hazard works gives a detailed account of their Civil War plant.

At Hazardville, near Hartford Conn., are the extensive Gunpowder Mills of the Hazard Powder Company, who have mills also in the towns of East Hartford and Canton. This company has eighteen sets of rolling mills with thirty-six iron wheels, each weighing eight tons; seven granulating mills; five screw press buildings; and three hydraulic presses of five hundred tons each, in different and separate buildings; and about fifty buildings used for dusting, assorting, drying, mixing, pulverizing, glazing, and packing houses—with extensive saltpeter refineries and magazines, cooper, iron, woodworking and machine shops—in all about 125 buildings at their main works at Hazardville and Scitico, extending over a mile in length and half a mile in width. To propel this vast amount of machinery twenty-five water wheels and three steam engines are required. This company manufactures annually over a million dollars worth of powder of various kinds, known as Government, Sporting, Shipping, and Mining Powder, of which large quantities were at one time exported to Europe. In the census of 1860 the value of the company's output is given as $991,500. (3)

Shortly after the war began, Ordnance Departments in both the Union and Confederacy were faced with problems regarding their gunpowder supply. The difficulties encountered by the South differed greatly from those of the North, and for this reason it is necessary to deal separately with the obstacles confronting each side.

Now what powder problems could possibly have plagued the Union at a time when they possessed two of the largest powder plants in the world? The Brandywine Mills were located just outside of Wilmington, Delaware. In the preceding election this state had abounded with secessionist sentiment, and although the Legislature refused to vote for secession, neither did it affirm loyalty to the Union. (4) The roving bands of pro-Southern marauders, who were burning bridges along the Philadelphia, Wilmington, and Baltimore Railroad coupled with the threat of sabotage, caused the DuPonts to create a Home Guard defense and to take every precaution to exclude strangers from their premises. (5) The Governor of Delaware commissioned Henry DuPont Major General of all the State troops raised or to be raised. (6) Now that he was in command of the Delaware militia, the mills along the Brandywine were not as fully prone to secessionist attack or sabotage. (7)

During the initial stages of the war, Colonel Hazard was suspected of being disloyal to the Federal cause, which presented the Ordnance Department with another of its problems. [8] Because he had previously lived in the South, had entertained Jefferson Davis, and had openly attacked abolitionist activities, he was distrusted and continually watched. As it turned out, these suspicions proved to be absolutely groundless. [9] The location of the DuPont plant combined with Hazard's questionable background gave the government much to worry about with regards to their powder supply.

The powder problem in the south was simply that resources and powder works were almost non-existent. At the outbreak of the war there were only two mills operating in the seceding states, the larger of the two (The Sycamore Plant) being located near Nashville. [10] The census of 1860 listed its yearly output at only $25,000. (11) Production could in no way be increased here because the saltpeter preparation facilities were of limited capacity. [12] The smaller mill was located in Walhalla, South Carolina, having been built to furnish blasting powder for a tunnel building operation, but was too small to supply powder for military use. [13] Powder obtained by the Confederates through their capture of the arsenal at Norfolk, Virginia, combined with small amounts left over from the Mexican War, afforded them about 60,000 pounds. [14] A war could hardly begin with a supply of gunpowder leaving so much to be desired.

Two alternatives presented themselves to the Confederate Ordnance Department. They could purchase powder from unreliable blockade runners, at $3.00 per pound, or they could construct a powder works of their own. Since acquiring powder from blockade runners was costly and undependable, Josiah Gorgas, head of the Southern War Department Ordnance Bureau, placed Colonel George Washington Rains in charge of a gunpowder manufacturing project. [15]

Immediately Colonel Rains set out to find a location favorable to the large-scale manufacture of powder utilizing the incorporating process. He selected a site near Augusta, Georgia, primarily because of its central location, railroad facilities, and security from enemy attack. [16] It was situated beside an unused canal, which would provide water power, transportation, and water that was free from earthy salts. By employing an intricate system of short cuts, he was able to construct, outfit, and operate a plant that would produce 2,750,000lbs. of powder between April 1862, and April 1865. [17]

Two of the three essential components of gunpowder, sulphur and charcoal, were easily acquired, but saltpeter had to be purchased from abroad. Willow wood, for charcoal, was abundant; and a large quantity of sulphur, imported from Italy for the sugar industry, was available for Rains' use when he began his project. Later, sulphur was obtained from iron pyrites. [19] The acquisition of saltpeter from overseas presented a most difficult problem because:

> A rigorous blockade of the seaports of the South was immediately commenced, through which the principal ingredients of gunpowder (saltpeter) had to be largely sucked in. [20]

In November of 1861, the steamship <u>Tennessee</u> was chartered to sail to Liverpool to procure a supply of saltpeter, but it was blockaded at the mouth of the Mississippi. [21]

As the blockade was intensified, the South was compelled to create a domestic supply of saltpeter. A Niter and Mining Bureau was established to foster the extraction of nitrates from caverns in eastern Tennessee and northern Alabama. Upon Federal seizure of many of these caves, the Bureau decided to encourage private parties to construct niter pits (nitriaries) which would produce nitrates by decomposing vegetable matter. [22] Although 2,700,000 pounds of saltpeter slipped through the blockade, about 3,000,000 pounds was produced within Confederate boundaries. [23] Thus through the efforts of the blockade runners, the genius of Rains, and the utilization of domestically acquired nitrates, the army of the South was kept abundantly supplied with powder.

For Hazard and DuPont to remain loyal to the government cause meant they had to abandon all hopes of gaining any substantial profit. The rapidly increased demand for military powders compelled them to hurriedly expand their resources and to convert many of their buildings and much equipment to wartime production. (24) They were then faced with the risk that many of these additions and innovations would have to stand idle if the war should suddenly end. (25)

Manufacturing powder for government consumption caused these two giants to lose many of the markets they had previously enjoyed and also to suffer financially. Confederate privateers put an end to powder shipments by boat to California. (26) The miners and railroad builders in that area needed blasting powder, so consequently a group of California capitalists formed the California Powder Company to satisfy these needs. (27) Writing to one of his agents in May, 1861, Henry DuPont summed up both his own and Colonel Hazard's predicament quite nicely. "The extra demand for powder for war purposes will not equal the regular demand which would have existed had peace continued". (28) The government was not prompt in paying its bills, and when payments did finally arrive they were only partially made in cash. (29) Hazard wrote DuPont that:

> The government being indebted to us to a large amount, and not having paid anything for the past five months or more, I wrote to the Secretary of the Treasury laying the matter before him and today received his reply stating that our bills would be paid in one-third cash and two-thirds in certificates of indebtedness. (30)

The failure of the government to pay promptly caused the powder men to demand cash payments from their private customers. They were forced to ask these customers for cash because their own credit was as heavily strained due to the fact that the raw materials and labor necessary to the manufacture of gunpowder required cash payments. Also, with the war becoming more intense, they feared many of their patrons to whom they had been extending credit would collapse financially. Here is what the Colonel had to say regarding an instance of this situation as it existed early in the war.

> *"I desire to say a word or two to you in relation to the business we are both doing in supplying powder to the copper mines in Lake Superior. You know we have both been in a measure forced into not only selling powder in that quarter at little or no profit, but into giving credit we ought not to have given because copper is a cash article and these miners ought to pay cash for their powder....Nothing but disaster is likely to come upon the country and plenty of users of our concerns who are good now will become bankrupted during the next six months."* (31)

Saltpeter again caused much concern, as it had to be secured from India, a part of the British Empire, which recognized the Confederacy as a belligerent. Maintaining harmonious relations with England was difficult because of the affair involving the British steamship Trent, which infringed upon British rights as a neutral. As a result, England prevented the shipping of saltpeter from her dominions.

> ".....we therefore, by and with the advice of our Privey Council, and by this our Royal Proclamation do order and direct that, from and after the date here of, all gunpowder, saltpeter, nitrate of soda, and brimstone, shall be, and the same are, hereby prohibited either to be exported from the United Kingdom or carried coastwise." (32)

The United States and Britain finally agreed upon terms acceptable to each other, and England allowed saltpeter to be transported to America once again. (33) Costs of this essential raw material continually spiraled upward because the supply in India was rapidly diminishing. (34) As a result, explosives prices were constantly pushed to higher levels in the entire industry. The following statement appeared annexed to price lists forwarded to Hazard agents in various Union cities.

> Since the issuing of our last price list of the 1st of September, 1862, in consequence of the great and unprecedented advance of foreign exchange and other causes, saltpeter, like all other articles has greatly advanced in price, and we are, in common with other manufacturers, compelled to advance the price of powder in conformity to the list of prices we hand you herewith. (35)

To further illustrate the retail price change evolving during the course of the war is a portion of two price lists, the first dated December 20, 1861, and the other August 1, 1864

CANISTERS		Dec., 1861	Aug., 1864
Electric Gunpowder	in canisters of 1 lb. each	.85	1.15
American Sporting	" " " " " "	.50	.85
Duck Shooting	" " " " " "	.58	.85
Indian Rifle	" " " " " "	.37	.70
Kentucky Rifle	" " " " " "	.40	.65
Fair Lawn Mills	in 1 lb. oval canisters	.30	.55

During the war many small companies came into existence to supply demands in their immediate localities. They were able to survive because the larger companies were, for the most part, busy filling orders for military consumption. These small concerns could sell their powder for much less than could the giants in the business. They had no transportation costs or problems to contend with and had not provided for explosions, as they had not been operating long enough to have experienced any such unpleasant mishaps. Consequently, when the war ended, the powder industry was faced with a condition of unrestrained competition resulting in both wholesale price-cutting and a very unstable market.

To make matters worse, the government auctioned publicly its war surplus of powder which sold at rock bottom prices. In April of 1866, Henry DuPont wrote: "The policy pursued by the Nave Bureau in crowding immense quantities of powder, by auction, on the market has completely broken down the trade." (36)

Much of the powder disposed of in this fashion was of superior quality, but some that came from dismantled ammunition contained much foreign matter. Both the Hazard Powder Company and the DuPonts bought large amounts of this powder which had to be reworked before it could be sold. In 1866 Mr. Pope, Secretary of the Hazard Company, informed a DuPont agent that the Hazard's had recently purchased government powder and that "it was found to contain percussion caps, nails, stones chippings of land and pieces of lead from inside shells. The workmen were so fearful of it that they dumped it all into the river, which is what the government ought to do with the whole of it". (37) As a result of the competition encountered from smaller concerns and because of the government flooding of the market with inexpensive and dangerous powder, the industry staggered and, as a means to survival, was forced to consolidate into a mammoth syndicate.

Footnotes – Chapter Three

1. Hazard to DuPont, New York, 22 Apr. 1861
2. Bessie G. DuPont, Manuscript Notes, quoting a company letter of 24 May 1861.
3. J. Leander Bishop, A History of American Manufacturers from 1608-1860. Vol. III (Phildelphia: Edward Young and Co., 1868), pp. 422-3.
4. Dutton, op. dit., pp. 92-3.
5. Norman B. Wildinson, "Operation Sabotage". The Civil War Times, Vol. II, No. 3 (E. J. Stackpole: Gettysburg, Pa., June, 1960). Pp. 6-7.

6. Dutton, op. cit., p. 93. Like Hazard, General Henry had been a Whig, and had been a delegate to their convention in 1844, which nominated Clay for the presidency. "Henry Clay was his political idol, a statesman, in his opinion, second only to Washington, and he smoked only 'Henry Clay' brand cigars." Dutton, op.cit., pg.16
7. However, during the Civil War period the Brandywine mills experienced ten major explosions which killed 41 men. Perhaps the Confederates did succeed in sabotaging the DuPont plant after all. Wilkinson, op. cit., p. 8.
8. Even though Hazard was thought to be disloyal, the Ordnance Department still saw fit to place orders as early as April 22, 1861.
9. Van Gelder and Schlatter, op. cit., p. 258.
10. Both these mills were stamp mills not capable of large-scale production.
11. Van Gelder and Schlatter, op., cit., p. 107.
12. Colonel George Washington Rains, History of the Confederate Powder Works (Newburgh, N.Y.: Newburgh Daily News Printers, 1882), p.5.
13. Van Gelder and Schlatter, 1oo. cit.
14. IBID., Wilkinson, op. cit., p. 5. Says they also seized 650,000 pounds from DuPont agents in the South at the beginning of the war.
15. E. Horton Coulter, "The Confederate States of America, 1861-1865", A History of the South, Vol. VII (Louisiana State University Press, 1950), p.205.
16. Rains, op. cit., p. 4.
17. Van Gelder and Schlatter, op. cit., p. 113.
18, Coulter, 1oo. cit.
19. IBID
20. "The Gunpowder of the Confederates", The Scientific American, Vol. VIII, No. 18 (New York: Munn & Co., 2 May 1883), p. 275.
21. Rains, op. cit., p. 6.
22. James Ford Rhodes, History of the United States from the Compromise of 1850 to the Final Restoration of Home Rule in the South, Vol. V 1864-6 (New York: Macmillian Co., 1907), p. 394.
23. Coulter, 1oo, cit.
24. Aside from being called upon to produce ordinary musket and cannon powders, Hazard and DuPont were requested to manufacture a very special type of cannon powder commonly referred to as "Mammoth Powder". The large bored Rodman and Dahlgren guns were bursting at their breaches because they would not resist the pressures created by the explosions of the big powder charges required for long range firing. Captain Rodman was able to measure these pressures and found that they could be modified by increasing the size and density of the powder grains. With this data Lammot then developed grains as large as three inches in diameter which, because they burned slower than ordinary cannon powders, would create a propelling instead of a blasting force. "Mammoth Powder" was to prove its worth in the epic battle between the Monitor and the Merrimac. Judging from a Hazard to DuPont letter written on Jan. 22, 1864, the manufacture of these large grains was difficult for them to continue. "This mammoth powder with Major Laidley's requirements for the size of the grain is a perfect nuisance to make and I think it very likely that it is not ordered from anyone but you and us and we both ought to charge fully 2 cents a pound more for making this powder or obtain some alteration of the size".
25. Dutton, op. cit., p. 92.
26. Van Gelder and Schlatter, op. cit., p. 119. Many new mills were started to furnish powder to markets in their immediate vicinity.
27. IBID., op. cit., p. 120. A Mr. Dutton and Abraham Manseau were employed by the California Powder Works and had received their powder making experience at Hazardville.
28. IBID.
29. IBID., p. 119. They say here the latter difficulties encountered by the hazard Powder Company (q.v.) had their inception during the period.
30. Hazard to DuPont, Hazardville, 22 Sept. 1861
31. Same to same, Hazardville, 24 May 1861.
32. Great Britain, A Proclamation by the Queen, 20 Nov. 1861, British Foreign and State Papers, Vol. LI, 1860-1 (London: William Ridgeway, 1860), p. 170.
33. Van Gelder and Schlatter, op. cit., p. 118.
34. By 1863, according to Van Gelder and Schlatter, a method was developed to convert the somewhat less expensive stores of Chile Saltpeter to render it satisfactory for use in the manufacture of military powder. This greatly relieved the demand for India's saltpeter deposits. To refine this South American saltpeter, a large soda refinery was constructed at Hazardville in the early part of 1864.
35. Hazard Powder Company, General Price List (New York: 11 Feb. 1863).
36. Bessie G. DePont, E. J. DuPont de Numours and Company, a History, 1802-1902 (Boston and New York: Houghton Mifflin Co., 1920), pp. 102-3.
37. IBID., p. 105.

INDIAN RIFLE GUNPOWDER
Manufactured By The
HAZARD POWDER COMPANY
New York. U.S.A.
Trade Mark Registered

PATENT CARTRIDGE

FOR

PARROTT'S

200 Pounder Rifled Cannon

Patented March 18th, 1862.

MANUFACTURED BY THE

HAZARD POWDER COMPANY,

HAZARDVILLE, CONN.

DIRECTIONS.

The hemispherical end of the Cartridge corresponds in shape to the bottom of the bore of the Gun. In loading, this end is first inserted into the Gun, and may be followed with the shot, and both sent home together.

Purchase of the Hazard Powder Company by DuPont and The Powder Trust Chapter Four

A Thesis by David N. Jellison

When Colonel Hazard died in May, 1868, he left his powder works in the hands of two elderly men and a third who was in poor health. (1) A few months later a powder agent in New York wrote:

> *The death of Colonel Hazard has effected the standing of their company in the trade very seriously. It is reported that the Company is falling to pieces. I think they are badly in want of business ability.* (2)

There was no one capable of taking Hazard's place, and the succession situation he had feared when his son, Horace, and John Garische were killed was now causing the company to flounder. Mr. Peter Cornell, who held one-third of the firm's stock, became President, even though he was an old man. Thomas S. Pope was made the company secretary and their general agent in New York, and A. E. Douglass was appointed treasurer.

The post-Civil War United States lapsed into an unstable period, which fluctuated between depression and prosperity and reached its climax with the panic of 1873. The powder industry, like so many others, was seriously affected by these unstable conditions which only served to further complicate the duties of Colonel Hazard's successors. John L. Winkler's DuPont Dynasty presents a lucid summary of the powder maker's appalling situation during this period.

> For five years the powder trade was about as tranquil as a dog pit. Agents stole customers from beneath each other's noses. Representatives of the same company undercut each other. Bribery and sabotage were common. The railroads openly gave rebates to concerns that furnished powder for their construction work. (3)

To avert ruinous price competition, to regain lost markets, and protect the industry from injurious legislation, a consolidation of some sort was necessary. (4) Consequently, the morning of April 23, 1872, saw delegates representing the principal American powder concerns congregate at Number 70 Wall Street in New York, where the offices of DuPont's general agent, P. L. Kneeland, were located.(5) A. E. Douglass, the secretary of the meeting, presented a plan for association known as the "Articles of Association". After some minor revisions, they decided to have the "Articles" printed and distributed to participating concerns, and then a committee of four was appointed to draw up price schedules for the major American markets. The "Articles" provided that the combination be called the "Gunpowder Trade Association of the United States", established for the purpose of guaranteeing "an equitable adjustment of prices and terms for sales of powder throughout the United States". (6) Firms were allotted votes according to their size; DuPont, Laflin and Rand, and Hazard were given ten each, the Oriental Mills six, and the others four. "With competition eliminated among the kingpins, the lesser fry would now be easier to handle". (7)

In 1871 a severe explosion ruined many of the mills and buildings at Hazardville which seriously effected the company's credit standing. Then the panic of 1873 came as a further setback to the stockholders. It was thought financial recovery could be gained if a member of their New York staff came to Hazardville to actively manage the plant. Thomas S. Pope was transferred to the works, but his lack of practical powder making experience and failing health seemed only to increase the troubles of the company. "In 1874 they were said to be in need of both money and orders". [8] By 1876 a majority of the Hazard stockholders, who were both old and discouraged, were glad to sell their holdings to the DuPont Company. This was the largest powder transaction to take place in the country up to that time. [9]

Although the Hazard Company was owned outright by the DuPonts, nothing was publicly known of the sale for at least ten years. [10] This condition existed because Henry DuPont allowed the Hazard to conduct its business as a separate corporate entity. [11] Aside from giving the DuPonts control of twenty votes in the Gunpowder Trade Association, this secret purchase greatly augmented their holdings in smaller powder concerns. In many cases the combining of Hazard and DuPont minority interests constituted a majority of a particular company's stock. Thus through the acquisition of Hazard, Henry DuPont now directed the affairs of many of the lesser firms in the powder industry.

The decade between 1880 and 1890 saw companies that manufactured 95% of the country's output of rifle powder and 90% of its blasting powder abiding by the Association's rules. When a new firm entered the industry they were usually politely asked to split a majority of their stock between the Laflin and Rand Powder Company, the DuPonts, and the Hazards. If an agreement was not reached, and the new concern sold their powder for less than the prices stipulated by the Association, it was not long before a course of unpleasant action was pursued by the combination. They would undercut the new firm's prices by at least 10 cents, bribe railroad agents to learn where the young company was shipping its powder, or establish a supply company in the area to undersell the new competitor. [12] If these measures failed, the Association would then secure high freight rates out of the vicinity where a new firm was operating. [13] In one such instance someone remarked, "It was perhaps a year until they died."[14] These procedures proved highly effective and are somewhat reminiscent of those used by the Standard Oil Trust during its ugliest years. [15]

The leaders in the black powder industry soon realized that much competition was developing from the increasing use of high explosives. In 1881 William Colvin, the treasurer of the Hazard Company wrote:

> The decline in price of High Explosives has already had the effect of increasing their use. And as they are considered four times the strength of black powder they naturally displace so much of the latter. [16]

If the black powdermen were to dominate the explosives trade in the United States, it was now mandatory for them to control high explosives also. The powder giants were forced to either build dynamite plants of their own or to become substantially interested in firms already operating. [17] By 1895, most of the American high explosives establishments were consolidated into one large holding company, in which DuPont, Laflin and Rand, and Hazard held a majority of the stock. [18]

The Hazard Powder Company enjoyed its peak production during the ten-year period between 1880 and 1890 when they manufactured 6,250,000 pounds of blasting powder and 2,500,000 pounds of sporting powder annually. [19] By the middle of the 1890's, the output of powder at Hazardville had declined rapidly. The greater portion of the blasting powder markets were located in the West. Railroad rates for the finished powder had been raised to a point where it had become necessary to discontinue the large-scale manufacture of blasting powder at the Hazard works. The freight bills for their saltpeter and brimstone had also been increased. Much of the demand for Hazard Black Sporting Powders had been displaced by the introduction of "smokeless" sporting powders. Although the Hazard Company sold a "Hazard Smokeless," it was manufactured for them by the DuPonts. [20]

The DuPont Company was responsible for the curbing of production of Hazardville. Since DuPont owned or controlled many of the powder factories in the areas where powder was used, they saw fit to end large-scale manufacture at their Hazard plant in order to eliminate the extra miles of shipping both raw materials and finished powder. A letter written in 1896, from Edward Prickett, superintendent at Hazardville, to F. C. DuPont, refers to an instance of this situation.

> *"It has been a matter of surprise to me that improvements have not been introduced here. From time to time I have expected to receive instructions to make changes and I have even suggested to Mr. Colvin, Pres., that we introduce some more modern way of doing some things, but have generally been met with what amounts to denial. The reply has been in substance, keep your plant in order but if we spend money on improvements it must be by way of a new plant out West."* [(21)]

In 1902, the leadership of the DuPont Company fell into the hands of three capable cousins. Alfred I. DuPont was a black powder expert, Pierre S. DuPont was an able financier, and Thomas Coleman DuPont was an excellent organizer. [(22)] Through their joint holdings with Hazard, they owned six plants outright, held substantial minority shares in sixteen companies, and majority interests in two. Of eighteen concerns in which the cousins held sizable interests, the Laflin and Rand Powder Company, whose management was old, also owned substantial titles in thirteen, controlled two, and had large holdings in two more. From a percentage standpoint, the DuPonts dominated 40% of their holdings, while leaving 60% of their interest unprotected if Laflin and Rand should suddenly decide to become non-cooperative or to sell out. The combined holdings of Hazard, DuPont, and Laflin and Rand controlled fifteen of twenty-two prominent explosives concerns and also gave them large minority titles in the remaining seven. By dominating these fifteen larger companies and by direct investment, the "Big Three" also held interests in over fifty smaller firms.

The DuPonts, fearing that 60% of their holdings were imperiled because of a rumored Laflin and Rand sellout, endeavored to purchase Laflin and Rand, and, in October, 1902, succeeded in buying a majority of its stock. The DuPont Company was now well on its way toward monopoly and thus saw fit to discontinue the Gunpowder Trade Association.

To build his trust, Coleman conceived a plan magnificent in its simplicity: Through a few operating corporations, the fewer the better, he would take over all the companies affiliated with the Gunpowder Trade Association; run their plants through his operating corporations; and, finally bind these operating corporations into one great holding company. This super-company would be able to fix prices and prescribe policies with a great deal more certainty than a trade association. [(23)]

By 1906, the DuPonts were selling over 70% of the explosives consumed in the United States. [(24)] They owned stocks in more than one hundred concerns of which sixty-four had already been dissolved. [(25)] If Coleman had succeeded in dissolving them all, after having purchased their property outright, the title to it probably could not have been impaired by the government. [(26)] On July 30, 1907, in the United States Circuit Court for the District of Delaware, the Federal Government filed suit against the DuPont Company for violation of the Sherman Act. [(27)] The trial dragged on for almost five years before the court finally handed down a verdict which found the defendants guilty of restraining commerce in explosives "and with attempting to monopolize and monopolizing a part of such commerce". [(28)] The combination was enjoined and offered to dissolve because original conditions in the industry were impossible to restore, the court was forced to ask the DuPont Company to assist in the dictation of their own punishment. Working together with the court, they split the explosives business into three firms, the Hercules Powder Company, the Atlas Powder Company, and the E. I DuPont de Nemours and Company. Rules were then formulated to prevent a recombining of these interests. [(29)]

The Hazard mills were transferred to the Hercules Powder Company on December 15, 1912. On January 14, 1913, a severe explosion occurred, which damaged the plant to the extent that the new management decided to shut it down. Rather than rebuild at Hazardville, they moved the equipment to their works at Valley Falls, New York. Thus seventy-eight years of powder making beside the Scantic came to an end.

Footnotes - Chapter Four

1. Dutton, op. cit., p. 121.
2. Bessie G. DuPont, op. cit., pp. 109-10.
3. John K. Winkler, The DuPont Dynasty (New York: Reynal and Hitchcock, 1935), p.122.
4. In 1868 H. M. Boies, afterward president of the Laflin and Rand Company, drew up a plan for a gunpowder trade association, Instances of ruinous legislation stemmed from tariffs on saltpeter and brimstone which, according to a letter written from A. E. Douglass to the DuPonts on 20 May 1972, were influenced by the personal interests of some of the Congressmen. The powder industry also suffered from storage law problems which were often place in the hands of petty state and local officials.
5. Those present at the meeting were: Henry DuPont and F. L. Kneeland of the DuPont Company, A. E. Douglass and Thomas E. Pope of the Hazard Powder Compan, J. M Boies and T. C. Brainard of the Laflin and Rand concern, A. G. Fay of the American Powder Company, Arthur Williams and J. M. C. Armsby representing the Oriental Powder Company of Maine, and A. G. Fay of the Miami Powder Company. The Austin Powder Company was unrepresented, but through a letter expressed "its concurrence in the object and intentions of the meeting".
6. William H. S. Stevens, "The Powder Trust, 1872-1912," The Quarterly Journal of Economics, Vol. XXVI (Cambridge, Mass: Harvard University Press, Nov 1911), p. 446.
7. Marquis James, Alfred I. DuPont: The Family Rebel (New York: Bobbs-Merrill Co., 1941), p. 73.
8. Bessie G. DuPont, op. cit., p. 110.
9. Dutton, op. cit., p.110
10. Van Gelder and Schlatter, op. cit., p. 262.
11. By maintaining a separate existence Henry DuPont was not faced with the problems of administering two huge powder concerns, and, by keeping the transaction a secret, all the small companies purchased in the name of Hazard in reality would belong to the DuPonts.
12. Stevens, op. cit., p. 458.
13. IBID.
14. IBID
15. Winkler, op. cit., p. 143. The association also stifled competition by paying individuals not to enter the powder business.
16. William S. Colvin, Denver, Colorado, 11 June 1881, letter to R. L. Wheeler, New York, File Box 12, Lammot DuPont Collection of Correspondence, uncatalogued, Hagley Museum Library Foundation.
17. The DuPonts built a dynamite plant on Repauno Creek in New Jersey, and Laflin and Rand had large interests in the Jasper and Rand Dynamite Compan y. Hazard, Laflin, and DuPont each held one-third interests in the Atlantic Giant Dynamite Company, and in 1882 reorganized it into the Atlantic Dynamite Company. By this time, the DuPont's also controlled the Hercules dynamite concern.
18. The DuPonts, Laflins, and Hazards combined all their dynamite interests into the Eastern Dynamite Company in 1895, which was a large holding company dominating most of the industry. The concerns they did not control found it expedient to make trade agreements with the Eastern Dynamite Company. The "Memorandum of Understanding", entered into between the Eastern and Aetna Companies in 1895, illustrates an agreement of this type.
19. Van Gelder and Schlatter, loo. cit.
20. IBID.
21. Edward Prickett, Hazardville, Conn., 28 Aug. 1896, letter to Francis G. DuPont, Wilmington, Del., File Box 8, Francis G. DuPont Papers, Folder 2, Hagley Museum Library Foundation. In 1898 the DuPonts had drawn up plans for a Hazard Company to be constructed at Gauley Bridge in West Virginia on the Chesapeake and Ohio Railroad. This was to supply coal mines and to stifle competition in the area.
22. Thomas Coleman DuPont will hereafter be referred to as Coleman.
23. Winkler, op. cit., p. 161. In 1903, the E. I. DuPont de Nemours Powder Company of New Jersey was formed as the holding company "which Coleman had planned as the towering apex of his financial structure". Its capital stock was $50,000,000, half common and half preferred. The DuPonts held $15,000,000 worth of preferred and $13,000,000 of the common.
24. Dutton, op. cit., p. 194.

25. Coleman was also planning to discontinue the Hazard, Laflin and Rand, and Eastern Dynamite Companies when the courts intervened.
26. Stevens, op. cit., p. 480.
27. The case against them pointed out how for two generations the DuPonts and the Gunpowder Trade Association had conspired to control the explosives industry. They now dominated 70% of the trade which constituted a dominance of the industry and, at the same time, restrained interstate commerce. Because Coleman had not dissolved the Gunpowder Trade Association until after the purchase of Laflin and Rand, it appeared that the DuPont Company had been conspiring to rule the industry.
28. Stevens, loo. cit.
29. Stipulations were made to prevent the stocks of these three companies from falling into the hands of a voting trust. The defendants were enjoined from transferring their business to either of the two new concerns, from establishing trade of price agreements in localities or with customers, from waging a campaign of local price cutting to eliminate competition, from using the same office force as the two new companies, and from operating bogus independents. The three new firms were forbidden for five years to form any type of interlocking directorates, use the same sales agents, or create any new holding companies.

Letters from Colonel Hazard
And the Hazard Powder Company

The following letters are published here to show the importance of Colonel Hazard's status in the gunpowder industry, his association with the government regarding orders for gunpowder during the Civil War and his continued connection with the Duponts.

The Hazard Powder Company - one of the 'big three' which included Dupont, and Laflin & Rand - produced more than 40% of the gunpowder used by the Union forces during the war between the states.

Colonel Hazard's handwriting usually became more and more illegible as he continued writing, especially if he was agitated or concerned about a certain situation. He also tended to run sentences and/or thoughts together so as to make deciphering his handwriting a major task. The spelling of certain words was arbitrary at times and the letters are reproduced using the exact spelling and phrasing as in the written version. No attempt was made to correct any of the grammar or composition.

Subsequent letters, written after his death by officers of the company, are similarly difficult to decipher until, ultimately, they are typewritten.

Pictures and other visuals are used, intermittently, between letters to ease the readings and add interest in the subject.

Courtesy of Hagley Museum and Library

Enfield, April 10/1855

Gentlemen,

Yours of the 7th just was duly received, but remained unopened with many other business letters until today in consequence of the dreadful calamity which has befallen me by the sudden & appalling death of my son, of which this the public journals have doubtless informed you.

In reply to the subject of your note I would say that we shall not vary our terms to Mr. G. M. & Co. from 17, seventeen cents cash or 3 ½ percent interest added if they want six months.

Yours truly,

A. G. Hazard

To Mr. E. I. Dupont de Nemours & Co.

Wilmington

Courtesy of Hagley Museum and Library

New York, Feby. 23rd, 1860

Mr. E. I. DuPont deNemours & Co.
Wilmington, Delaware
Gents

Confirming our respects of yesterday we have nothing new to add today other then to say that Mr. Robert Williams who by their letter received this morning advise us that they are compelled to act as our brokers in the purchase of the Saltpetre of Goddards or they would not have been able to have secured the parcel it being by far as they say the best parcel in Boston and worth fully half cent a pound more than a parcel of 800 bags (which by the way he tried to get) bought by the Oriental Powder Co. at 9 ¼ cent same day. Hence Mr. Williams charge for purchasing is as usual one percent for buying shipping re the terms of sale are 9 ¼ cent less ½ percent 6 mos. or 4 ½ percent off for cash at our option. This condition we made ourselves as we happen to have some unappropriated money on hand. We shall therefore cash our half of the bill and you can do the same or give your note just as suits your conscience.

We don't suppose we shall be able to get any more Saltpetre of this quality at less then ten cents, tho we may get a small parcel of 500 bags, a very good parcel which we have wrote about if so we will advise you now as regards advancing the price of powder, our ideas are that it will not be best to advance the price of Sporting Powder at present if at all, now is it good policy for you and us to put up the price of Blasting Powder just at present until the small mills feel the effect of the advance in the price of Saltpetre say about 1st of April, <u>except</u> upon Lake Superior, we aught to write our respective agencies at once to advance the price there 25 cents a keg. If you approve this suggestion we will write our agents to this effect upon the very day & date you say you will. We aught also to advance the price of powder at Richmond & at Corington, Virginia. There is no sort of danger of any manufacturer's interfering with us at either of these points, so if you agree with us in this suggestion we will also give the necessary *instructions to our agents at Richmond & at Corington and about the first of April we think Blasting Powder can be put up all over twenty five cents a keg, but whether & when it is best for us to advance the price of Sporting Powder is a question that requires further and careful consideration. It is not for our mutual interest to encourage others into the manufacture of Sporting Powder so will reserve this for further consideration & conference with you. (over)*

Yours truly, A. G. Hazard, Prest.

P. S. Upon reflection in as much as we could not carry out our plan as we supposed we could & get 3 or 4000 bags & as it may harm some future joint operation to have Williams & Goddard know that we designed a joint a/c move we will send you the bill when we get it & you may send us your note for one half and we will settle for the whole & then they in Boston will know nothing about the concerted move.

Yours, A. G. H. Prest.

Mr. Dupont & Co.

I don't care about Williams knowing you have half but don't want Goddard to know we told Williams not to ship until we directed him.

A.G.H

Courtesy of Hagley Museum and Library

New York, July 24th, 1860

Mr. E. I. Dupont d Nemours & Co.,

Gents,

We have yours of the 23rd, and note its contents, in reply beg to say that we will accept your generous proposal and will take all of the Goddard lot, we will never again intrust to Mr. Williams & Son the execution of such operations had the writer been I Boston himself we should have got at least two & we think three thousand bags. I am now satisfied that Mr. Williams & Son who do the purchasing for all the small powder makers cannot do either you nor us justice in the execution of any large orders, because they sease upon the opportunity to protect & supply all the others at some time and when it is for your interest & ours to have the market advance it is policy for W & Son to keep it down as they are now trying to do

So in future we shall make our own purchases & not put it in their power to tell an importer who has 1200 bags that they will take it all & give them H. P. Co. note for 1050 & "Tom Jones" note for the balance, & to make Tom Jones or "W. Russell" who pays 10 or 20 percent for a discharge to purchase just as low as Dupont & Co., or H. P. Co., - this game shall now be blocked as far as we can do it and we hope you will do the same, as far as you can, tho all the orders we sent Mr. Prink go through Williams hands, we don't wish to harm Mr. Williams because we have always been his friend & have put a very large amount of business & commissions into his hands, but we feel very sure about his putting out of our way & into the hands of others some 10 to 1200 bags of s'petre we know, we should buy as soon as a dispatch could reach him, - & this he felt obliged to do to keep right with others, as he thinks.

Yours truly,

A.G. Hazard, Prest.

Courtesy of Hagley Museum and Library

Entrance to the Powder Mills
Sign reads: Powder Mills, No Admittance

MINING POWDER
MANUFACTURED AT
FF
HAZARDVILLE CONN.
LOW GLAZED

Hazardville May 24 1861

Mr. E. I. Dupont de Nemours & Co.

Gents,

I desire to say a word to you in relation to the business we are both doing in supplying powder to the Copper Miners upon Lake Superior – you know we have both been as I think in a measure forced into not only selling powder at (all things considered) into selling powder in that quarter at little or no profit, but into giving credits we never ought to have given. Copper is a cash article, and these miners ought to pay the cash for their powder or not ask for a credit of over 90 days besides this it is only those companies who have ample means that will now be good & it is a matter of doubt who have or has ample means to sustain themselves through the Present & prospective state of things – My idea is this that if you & us go on sending our usual supply of powder to Lake Superior this year and allowing our agents to sell it to the miners in general & open our usual credit - we shall loose much more money then we can possibly make as it is soon pretty clear & well settled that nothing but disaster is likely to come upon all business of the country - and plenty of men and concerns who are now good will be <u>bankrupted</u> during the next six months – what course then I submit to you, ought you & us (who have so long and I trust will continue to act harmoniously together) to persue in the conduct of our respective business upon Lake Superior.

I do not believe there will be much powder sent there except that which you & us send there & unless we adopt some plan & in harmony as to price & terms we will of course lose money in that trade now I submit what shall we do. I would propose that we sell only for cash except to some few companies & I am not prepared to name each of them to where we will sell upon a short credit. I desire your view upon this subject as soon as convenient. Major Harner who left here a day or two since told me he expected to visit you in a few days & very likely this letter will meet him there. The Major left in his room at my house a letter & paper of some importance to him & I have enclosed them to him hoping they may reach him at your place, but if otherwise will you do him & me the favor to forward the letter I here enclosed to him – where you think he will be most likely to get it.

Yours vy truly
A.G. Hazard, Prest

Note: What I here said to you of the dangers of crediting at Lake Superior will apply to all other places & we shall be hoping to cooperate with you as to terms of sale any & every where – for as I look at things few will be able to meet their payments six months hence if the present state of things are to last & I am sorry to say I think they will last and steadily grow worse.

A.G. H.

Courtesy of Hagley Museum and Library

Hazardville, June 3rd, 1861
Hy Dupont Eqr

Dear Sir,

With this I hand you copy of a letter rec'd from Commodore R. Aulick, acting Chief of Ordnance which will explain it self. I have so replied as to induce him I think to through in the bonds. I am surprised that he should have made such a condition & will advise you when I hear from him again.

I duly rec'd the letter from your firm about the 344 kegs & 174 bls prize powder to be sold at Pha ~ and desire to share it with you if it can be purchased at such a price as will pay us handsomely the fact is there will be much of this kind of condemned powder to be sold by Govt and we ought to get well paid for if you were to drop them a line & suggest at which .... we think we will, we can have no intercourse with them, so please give us your view upon this point.

Now as to California we may have to pay more freight & now more risk in shipping to California & ought to get higher prices there for Blasting powder & we have just written our agent there to confer with yours & try & put prices up & we have also told our agent we would write to you & request you to write your agent upon the subject if the suggestions met your view.

But again as to Govt powder we really think we ought to have 20 cents a pound for it but of course we will charge same as you do & be governed by your view upon this subject. We send you this by express which seems to be the surest mode of carrying letters at this time ~ please address us here as the winter will be here some week or more longer, Hoping soon to hear from you we are

Your vy truly,
A.G. Hazard, Prest.

Courtesy of Hagley Museum and Library

New York, Oct. 31/61

Mr. E. I. Dupont de Nemours & Co.

Gents,

I desire to see your Mr. Hy. Dupont or some one of your firm upon some business which I think will be of mutual importance. Will it suit your pleasure & convenience to meet me at the Continental Hotel, Pha. on Saturday of this week or on any other early day you may please to mention if so please telegraph Mr. Douglass or write me here & oblige.

Your obsvt.

A.G. Hazard, Prest.

Courtesy of Hagley Museum and Library

Three views of the Canister or Packing House

ENFIELD RIFLE POWDER
F,G
FAIR-LAWN MILLS

Hazardville, Dec. 2nd, 1861

Mr. E. I. Dupont de Nemours & Co.

Gents,

I am in receipt of your of the 28th …… & in reply will say that I will be happy to meet you at N. York on any day you may please to name if you will be so kind as to telegraph me directed to "Windsor Locks" Connt the day you will be there & inform Mr. Douglass also that we may notify Mr. Townsend & his friends to be on hand & I really think the sooner we arrange this matter the better as these gentlemen are getting very flattering accounts of the actions of their powder cartridge in the various kinds of rifle cannon in which it is being tried – when we meet we will talk over the kind of powder as by next time we shall have further accounts from Europe – as the future price of Spetre must depend very much upon what we hear in a short time from the other side. Hoping soon to hear from you I am

Your obsvt, A.G. Hazard, Prest.

N I sincerely regretted to hear of the sad accident at your mills, as I know by similar experience how to feel in such cases – but such belong to the dangerous & troublesome business in which we are both engaged & can't be helped or guarded against. I am, A. G. H

Courtesy of Hagley Museum and Library

Private & Confidential

Hazardville Jany 22/64

Mr. E. I . Dupont DeNemours & Co.

Gents

This Mammoth Powder with Majr. Laidleys requirement for the size of the grain is a perfect nuisance to make & I think it very likely that it is not ordered from anybody but you & us and we both ought to charge fully 2 cents a pound more for making this size powder or obtain some alteration of the size, that is have permission to put in some finer and some courser grains. We have had some of it rejected but we cannot make the powder & make a fair profit upon it at present price – now what say you to this proposition; shall we advance the price of it, or get if we can, an alteration in the size of the grain. We ought to do one or the other as very likely it will get more & more into use.

Yours truly

A. G. Hazard

Courtesy of Hagley Museum and Library

Hazardville Apl 18/64

Mr. E. I. Dupont Denemours & Co.

Gents

I thought I would say to you that I go to NY today and shall remain there nearly all of this week up to Friday & it may be we shall be so fortunate as to meet there. I hope it may so happen and can.

Your vy truly

A. G. Hazard

Courtesy of Hagley Museum and Library

Fire apparatus used by the Hazard Powder Company

FAIRLAWN GUNPOWDER
F.F.G.
MANUFACTURED
BY THE
HAZARD POWDER COMPANY.

Office of the Hazard Powder Co.,
89 Wall Street
New York

June 3d, 1864

Mess. E. I. DuPont de Nemours & Co.

Wilmington, Delaware

Dear Sir, Having changed the Secretaryship of our Company, we hand you herewith the signatures of the executive officers.

Very respectfully,

A. G. Hazard President

P. C. Cornell Vice-President

Tho. S. Pope Secretary

Courtesy of Hagley Museum and Library

Mr. E. I. Dupont DeNemours & Co. **Hazardville Sept 9, 1864**

Gents,

In the execution of orders for the Army we charge the cartage of the powder from the mills to the river about 3 ½ miles & then add the freight per vessel to New York & these charge the Gov't allow and pay you and other manufacturers do the same as we understand it but how is it now with the powder delivery to the Navy will you please inform me whether or not you receive from the Navy any compensation for the delivery of powder to the Navy to N.Y. Navel Arsenal for example or else we never have made any such charge to the Navy heretofore but now have agreed to execute an order for the Navy at 35 ½ cents upon conditions that they would allow us as the Army does the last of transportation – and in reply have just received a letter from Capt. Wise, Chief of Naval Ords. that the Bureau cannot take any responsibility in regard to the transportation or powder from our mills and can only order on the usual terms, that is we to deliver the powder to the New York Arsenal at our own cost - now how shall we meet and manage this matter. What shall we do submit and to work cheaper for the Navy than for the Army, or shall we charge the Navy one cent or half cent more per pound to cover these charges & cost of transportation – or how – please give me your view & what you will do as we doing & is going to do in this matter & oblige.

Your Obsvt, A. G. Hazard, Prest

Courtesy of Hagley Museum and Library

Private & Confidential

Hazardville Mar 13/65

Mr. E. I. Dupont Denemours & Co.

Gents

I have just received a letter from Capt. Wise wishing to know at what price we will charge per barrel for manufacturing domestic …. supplied to us by the Naval Bureau. Nitre to be delivered to our mills free of expense to us & I don't wish to give an answer until I can confer with you now what price if you please do you think it best to charge powder to be delivered. I suppose like all powder for the Navy at our expense will you please telegraph me by simply saying how many cents per pound as how many dollars per barrel direct the dispatch at "Thompsonville" Ct. and write also by mail. I suppose we are just to charge say $15. a barrel but will do as you think is best. Please reply at once.

Your truly

A. G. Hazard Prest

In your dispatch say twenty cents per pound or so much a barrel. I will understand the balance. A. G. H.

Hazardville May 2d, 1865

Mr. E. I. Dupont Denemours & Co.

Gents

Mr. Law has shown me the memorandum of a contract agreed upon between you & Mr. Green & Wilson & these which is all substantially correct & will inable us to proceed & I only want it so far altered as to protect us - I want the license made direct to us so that we can imploy Mr. Law or whom we please to make the kegs - tho we shall no doubt have Mr. Law make all the kegs we use as we have intire confidence in them as correct fair & honorable men but they may change their business so it seems proper that we should posess the right to have the kegs made by ourselves or by whom we please.

Yours Truly,

A. G. Hazard, Prest

Courtesy of Hagley Museum and Library

THE LINCOLN MEMORIAL

Enfield Apl 18th 1865

Gents,

The acting Secy of State at Washington has announced to the nation that the funeral ceremonies of Abm Lincoln the lamented President of the United States will take place at the Executive Mansion in Washington tomorrow the 19th just at 12 o'clock.

I request you to have our mills stopped and all work suspended that those imployed by us may have an opportunity of joining their fellow citizens in the observance of the day as one of national sorrow & mourning. Please give the proper notice and oblige.

Yours very truly,

Augustus G. Hazard, Prest.

of the Hazard Powder Co.

To Joseph Sexton

And

Edward Prickett, Esqr's.

Superintendent of the

Hazard Powder Co.

Hazardville, Connt

Courtesy of Hagley Museum and Library

Hazardville May 2d 1865

Mr. E. I. Dupont de Nemours & Co.

Gents

The Navy department informs me that they have a large quantity of condemned powder on hand at the N Y Arsenal also a quantity of compressed patent cannon cartridges also think to know upon what terms we will rework it or purchase it. I have replied by saying I would examine the powder and let them know what we would do. I have not yet examined this powder but Mr. Cornell informs me there is of it 3500 barrels already in the Arsenal & 2000 more expected in a few days & there is also about 500 boxes of various sized cannon cartridges from 2 to 10 pounds now I presume you have had a letter from the department making some inquiry & there will be no doubt a very large quantity of this kind of powder to be disposed both by the Navy & Army also and I have thought best to confer with you before making any proposition to the department as to what we would do and would be glad if you would give me your view as to whether or not it will be best to act in concert in this matter. We would prefer to as to making returns to the A. L. Assessor I have had an interview with that office from district here & his reply was that what had passed between us with respect to the old law about which I spoke to you when last we met did not now apply. He thought he said the new law gave us all the latitude we could ask as it permitted us to deduct 1 st freight from the place of manufacture to place of delivery. 2nd - "The reasonable commission net our 3 pct and other expenses of sale bonafide paid" no commission to be deducted when the article is sold at place of manufacture. So I suppose upon powder supplied to the Army we can make no deduction but must value it at the price we get from powder manufacturer for the Navy we can deduct I suppose only the cost of its delivery cartage freight & insurance – from mills to N Y or wherever delivered – then upon our main sales to the trade what can we and what shall we do – the law and our assessor here says we may deduct all charges actually paid & we have hithertoo taken our N Y list price for our standard & propose to do it again & for example – we sell a keg of FFF Gr. powder for $9 - at N Y less 5pct cash or in 6 mos time – now what does this powder sell for or what does it net us & so with all other kinds but we take the keg of FFF G. for example sold for $9.00

less 5 pct in is	.45 cents	
less freight & insurance to N Y say is	.14 cents	
cost of receiving & delivery at magazine at N Y say upon an average is	.50	
3 pct commission _	.27 cents	
	1.36	1.36
		7.64

Now what are your views upon this matter is this too much or too little to deduct make the deductions for receiving our delivering blasting powder 25 cent a keg if that be best. What say you to all of these things or I will meet you in N Y any day this week you please to telegraph me.

Yours truly

A. G. Hazard Pr.

N. I ought to have written you before on this subject but I have been away attending to the rebuilding our mill at Canton which was blown up while I was in Pha.

A. G. H.

Alf Blunden at Graining Mill – killed 1898

Narrow Gauge Railroad Track

SEA SHOOTING GUNPOWDER
Manufactured at
F
G
Hazardville, Con. by the
HAZARD POWDER CO.

My dear sir,

As we have the price of powder to fix – the Navy condemned powder to talk over and the iron keg agreement to fix value of taxed powder - I hope it may be convenient for you to meet me at N Y some day this week & if so please telegraph me at once at Thompsonville.

Yours
A. G. H.

To Mr. Hy. Dupont
May 2/65

Courtesy of Hagley Museum and Library

Confidential

New York May 16/65

My dear Sir

I have your letters & dispatches & regret that I cannot meet you here because it is very important that we personally confer or that you & Mr. Cornell personally confer together before we make an offer for so large a quantity of powder – apart of this powder I examined yesterday some 2500 bbls of it is now being landed & as all was not ready I wrote the department we would further examine it when the balance is landed & then make them an offer for the whole – it is not easy to make an examination of this powder – there is so much of it all piled up in one great pile & mass – the gunner?? Thinks there is from an 1/8th to ¼ of it considerably damaged & not too with saltpeter – much more of it is more or less harmed & still more of it in very good condition & not much harmed. It consists of your make ours. Oriental Schaghticoke Co. & Dover works or Union Powder Co. I believe they call it, mostly I should say of yours & our make – no one the gunner informs me from any quarter has been to examine this powder until I come there is a large quantity & it will not bring as much as you, I think from the tenor of your letter, suppose it will bring – we are dealing with a declining market & before either you or us would probably use it up saltpeter may be down to (10) ten or twelve cents per pound & my idea now is that we ought to purchase this powder at say $12 a barrel there being so large a lot – but as I said before there are some things in the matter we cannot so well understand each other by writing & I hope you will be able to come to N Y and see Mr. Cornell & you & him talk the matter over before we make an offer - I must return home tomorrow – tho I would come down again to meet you if that be necessary – but I think you & Mr. Cornell can settle all that is required to be understood in the matter.

Yours truly

A. G. Hazard

Hy. Dupont & Co
New York
May 16 1865
A. G. Hazard

Bureau of Ordnance

Navy Department

Washington City June 5, 1865

Mr. A. G. Hazard

President Hazard Powder Co.

Hazardville, Conn.

Sir:

Your letter of the 3d inst. in relation to the barrels containing the damaged powder at New York and Boston is received. The Bureau has no objection to your having all barrels which are found to be in the condition you represent, and has therefore directed on examination to be made by the Ordnance officers at Boston & New York for that purpose. But such as are found to be in sound and good condition, the bureau must retain for use in the transportation of Nitre. As soon as the reports alluded to are received, you will be duly notified.

I am Sir, Your Obedient Servant

Signed R. Aulick

Asst. Chief of Bureau

Bureau of Ordnance

Courtesy of Hagley Museum and Library

The building on the right in the upper photo is shown on the left in the lower photo. From left upper (which is approximately the office location) to lower right (which is approximately Raffia Road area) is a panoramic view of one expanse of the Hazard Powder Company. This photo was taken in the area of Wilstar Circle.

SHIPPING POWDER
MANUFACTURED BY THE
FF
HAZARD POWDER CO.
BRIGHT GLAZED

Hazardville June 7/65

Mr. E. I Dupont De Nemours & Co.

Wilmington, Delaware

Gents

I am in receipt of yours of the 5th and hand you here both copy of Cmd. Aulick's letter just rec'd. I did think he would not make a point about the bonds but we had as you suggest better yeald it and give him about half of them which we can arrange to do – the fact is the Navy Bureau has purchased a large quantity of saltpeter of the New Haven Chemical Works to be made up for them doing the present year & want these barrels to use at that point. So I will make no objections via rely to his (Aulick's letter) but wait his order for the delivery of the power.

Yours truly,

A. G. Hazard, Prest

N Y. There is no powder I believe at Boston Navy Yard to be purchased in this lot that we are aware of but some of the compressed cartridges are there & if there is any powder there – we have not seen it but stated in my proposal that we would take the few compressed cartridges that were there same as those at N Y & will send you copy of their order & report when we receive it.

A. G. H.

Courtesy of Hagley Museum and Library

Hazardville Jun 17/65

Mr. E. I. Dupont De Nemours & Co.

Gents

With this I hand you copy of letter recd from Chief of Naval Ordnance to which I have replied that we will take the compressed cartridges but in as much as they have reduced the damaged powder from over (3000) bbls to less than one that will require a reexamination as what is now left maybe with no more than the saltpetre that is in it which very likely is the case - tomorrow is next day & report again to them & to you.

Yours truly

A. G. Hazard

Courtesy of Hagley Museum and Library

Bureau of Ordnance

Navy Department
Washington City,
July 5, 1865

Mr. A. G. Hazard
Prest Hazard Powder Co.
Hazardville, Conn.

Sir,

The survey of the damaged powder and powder barrels at New York and Boston has been completed and the result is as follows:

	Barrels	Powder
At New York	999	99900 lbs
At Boston	29	2887 "
	1028	102787

Compressed powder

	Powder
At New York	48523 lbs
At Boston	720 "
	49243 lbs

Whenever you will notify the Bureau that you are ready to receive the above the order for the delivery will be given to the Ordnance Officers at Boston and New York.

I am, Sir, Your Obedient Servant
R. Aulick
Assis't Chief of Bureau

Courtesy of Hagley Museum and Library

Saltpetre Refinery

Office Complex

SHIPPING POWDER
MANUFACTURED BY THE
F
HAZARD POWDER CO.
BRIGHT GLAZED

SPECIAL MEETING
of the
BOARD OF DIRECTORS
of the
HAZARD POWDER COMPANY

The New York Office of the Hazard Powder Company
Edward Prickett is seated center – 3rd from left
Also pictured is Mr. Colvin on the right.

A. E. Douglass VP

Tho. L. Pope, Sec.

23 July 1868

A special meeting of the Board of Directors was held at their office in New York on Thursday 23 July 1868 at 10 A.M.

Present – A. E. Douglass V. Prest. In the Chair
Alex H. Bullock)
Peter C. Cornell) Directors
George L. Cornell)
& Tho. S. Pope - Secretary of the Company

The Minutes of the Regular Meeting of 3 March 1868 were read and approved.

The following Preamble and resolution were then presented by P. C. Cornell, seconded by Geo. L. Cornell and adopted.

Whereas this is the first meeting of the Board of Directors of The Hazard Powder co. since the death of their late President which occurred on 7th May last and

Whereas this Board share with his family the affliction of so great and grievous a loss and desire to express their sympathy with them in so sad a bereavement & testify their own regard for the memory of one so valued while living, Therefore

Resolved That this Board tender to the Family of Augustus G. Hazard late President of this company their earnest condolence and sympathy in the loss which they in common with the Company have sustained by his death, an event which has deprived this Corporation of one whose master mind originated and animated its business: to whose energy, business ---- and --- its present pre-eminence is mainly due and who had by great & distinguished personal qualities endeared himself to his associates.

Resolved that the Secretary be directed to transmit to Mrs. Hazard & family a copy of the foregoing Preamble and Resolutions

The Vice President and Secretary reported that they had submitted to the President of the Company the resolution passed at the meeting of this Board held 3d March 1868 directing that the amount then standing to the Credit of Surplus "be transferred to the credit of an account to be opened on the Company's Books and styled "Construction" subject to the concurrence of the President – and that he had concurred therein.

On motion of P. C. Cornell seconded by Alex. H. Bullock it was

Resolved that the Vice President and the Secretary be authorized to dispose of the Canton Mill Property at a price not less than Ten thousand dollars ($10,000) reserving to the Company all the machinery and inserting in the deed of sale a provision that it should not be used for the manufacture of Powder.

On motion of P. C. Cornell seconded by G. L. Cornell it was

Resolved that this Board gives its consent to the transfer by A. E. Douglass of one share of the Capital Stock of this Company held by him to Thos. L. Pope.

The Secretary was directed to notify the stockholders of this Company that the Annual Meeting of the stockholders of the Hazard Powder Company would be held at the Office of the Company in the town of Enfield, Connecticut on Tuesday 4th August 1868, at noon for the purpose of electing five directors for the ensuing year and on motion of Geo. L. Cornell seconded by Alex. H. Bullock it was

Resolved that Peter C. Cornell)
Edward Prickett) be appointed Inspectors
A. D. Bridge) of said election

On motion the Meeting adjourned

Tho. L. Pope, Sec.

Courtesy of Hagley Museum and Library

Colonel Hazard is buried in the Enfield Street Cemetery

August 1868

A regular meeting of the Board of Directors of The Hazard Powder Co. was held at the office of the Company in the town of Enfield, Connecticut on Tuesday, 4th August, 1868 at 4 P. M.

Present – Alex. H. Bullock, Peter C. Cornell, Geo. L. Cornell and A. E. Douglass, Directors. On motion Alex. H. Bullock, was called to the Chair and A. E. Douglass was appointed Secretary, pro tem.

The following communication was received and ordered on record, to wit:

Enfield Conn. Aug. 4, 1868

We the subscribers appointed Inspectors of the Annual Election for Five Directors of The Hazard Powder Company to be held at their office at Enfield, Connecticut this 4th day of August, 1868, do hereby certify that five thousand and seven votes have been cast, and for the following persons, each having received the five thousand and seven votes cast

Alexander H. Bullock

George L. Cornell

Andrew E. Douglass

Thomas L. Pope

Peter C. Cornell

Topeka, Kan June 11, 1881

R. L. Wheeler Esq. V. Pr.

Hazard Powder Co.
88 Wall St. N. Y.

Dear Sir,

Replying to your favor of the 6th – I had previously been advised of the shipment made by Mess. E. C. Meacham & Co., St. Louis. The shipment went via the M. K. & T. to Junction City and thence via the Union Pacific to Denver, and was evidently smuggled through with the supposition on the part of the Railroad Companies that is was Baking Powder. I am investigating the matter however and think that parties are sufficiently advised upon the subject to be on the lookout for future shipments of this character of freight under a false name. It seems that the powder was shipped without the proper brands upon the boxes as required by law. Should any future shipment be offered in this condition we shall certainly refuse it, at any rate until the law is complied with. We shall furthermore be on the lookout for shipments of fuse under the name of hardware. So far as there being any cut rate made over our line on powder I can assure you that we have quoted nothing less than tariff rates to any one. If shipments are made to western Kansas points and re-shipped from there they pay us the sum of our local rates to the Western Kansas points, and thence to Peublo besides paying the Denver & Rio Grande local from there to Denver provided that be the destination. I am not aware however that any shipment has been made in this manner. I replied to your telegram in substance as above. I cannot see where any advantage could come to the shipper by paying our local rate to Western Kansas points, and then again paying local from there; as the sum of the two would greatly exceed the through-rate from the Mo. River.

Yours truly,

(signed) J. F. Goddard, G.F.A.

Courtesy of Hagley Museum and Library

Scene from Powder Hollow and the Hazard Powder Company

Denver, Colo. 11 June 1881

R. L. Wheeler Esq. V. P.

Dear Sir,

This in reply to so much of yours of 6th inst. as has not been anticipated in letters since received by you.

The decline in price of High Explosives has already had the effect of increasing their use and as they are considered four times the strength of black powder they naturally displace so much of the latter. The Contractors opposing us, are just now burning, as little black powder as possible even though it be cheaper for certain work. They say that at present prices of H. E. they can get along without black, yet take pains every time they come to Denver to call & repeat, that so soon as we come to their terms viz $3.50 per keg they will return to us.

Freight for California, I gave you in a letter written yesterday in detail. $900.00 for 750 kegs, equal per keg $1.20. The Cal a Co. have had so little trade that desk room in a clothing store is the only office required their teaming being done by transfer wagon & City Express.

Receiving, per Keg	.03
Delivering per Keg	.10

Other expenses such as Salary re at present very small, as Mr. Larned has no help of any kind. Hazard Powder is conceded by all to be the most reliable and the popularity of Mr. Raynolds is without question your instructions to "stand firm on prices and await results" will be obeyed. I am not aware that any "irresponsible" Contractors have joined the Syndicate "so called". We are selling none who are not in good repute.

Spiegelberg Bros. Albuquerque, as reported yesterday have agreed to take a car of Atlas, at 24 cents less 5% frt added, 90 days. The 5% off from 24 cents making the price less than the last reduction to 23 cents. I did not advise them of the latter. E. C. Meacham & Co. of them I have already advised you fully. I note what you say as to the Kansas Agency nearest here, being Larned $3.50
(carloads $3.40). From there at that figure the present Colo. Price could be cut, if allowed,

J. F. Goddard G. F. Agt. To him I wrote fully regarding Meacham's shipment of "Baking Powder" but without giving name. He replies that he has written Mr. Vining fully, and we may rest assured that no further shipments will come over the U. P. or the A. T & S. F. at less than regular rates. The U. P. office here have also advised Mr. Vining & the pool Commissioner here whose duty it is to report such matters: "Judson Powder:" the car reported in yesterdays letter proves to be "Judson" consigned to Rowarth. 400 Cases each 50 lb. 20,000 lbs net frt $2100.00 packed in tight boxes lined with 3 or 4 sheets of paper. The powder resembles black Atlas somewhat, is darker in color, shines like Anthricite Coal dust, which it resembles in grain, which is soft, irregular, and about half dust. The Giant co. say this car will be well distributed, if it gives satisfaction, & the freight excludes its being brought from California a factory will be started here or further East, as was done in the case of "Giant" & "Hercules". Later information: The carload at the Depot is marked

Judson Powder Pat. Oct. 13 – 1876

Manufactured By the Giant Powder Co. office 210 Front St.Keep DrySan Francisco 50 lbs.

400 Cases each 59 lbs,
Freight to Denver $2180.50
Or 0.10 9/10 cents per pound

Am endeavoring to get the rate of freight over each road, viz. Southern Pacific, to Deming, N.M.

A.T. & S.F. to Pueblo
D. & R.G. to Denver

Your Obedient Servant
(signed) W. S. Colvin

Courtesy of Hagley Museum and Library

Powder Packed at Mills, Five Weeks ending June 30th, 1888

Reduced to Kegs 25# each

1888	Rifle	Shipping	Soda
June 2nd	42	2035	5029
9th		1901	5340
16th	1398	421	5695
23rd	531		5287
30th			4329
	1971	4357	25710

Total 32038 Kegs

Refined	month	ending	June 30th 1888.	Saltpetre 797 bags
do	"	"	"	Soda 1172 "
On hand	"	"	"	Saltpetre 649 "
Do	"	"	"	Soda 3542 "

E&OE New York, July 6th, 1888
The Hazard Powder Co.
Pr. Tho. L. Lequin

Courtesy of Hagley Museum and Library

Powder Packed at Mills. Four Weeks ending July 28th, 1888

Reduced to Kegs 25lbs. each

1888	Rifle	Shipping	Soda
July 7	"		1136
14	1539		3604
21	1003		3400
28	947		3685
	3489		11825

Total 15314 Kegs

Shipped to Mills during	month	ending July 28th/88			
				Soda	2150 Bags
On hand	"	"	"	Saltpetre	1559 "
do	"	"	"	Soda	4250 "
Refined	"	"	"	Saltpetre	313 "
do	"	"	"	Soda	1442 "

E&OE

New York July 31, 1888

The Hazard Powder Co.

pr. Tho. L. Lequin

Powder Packed at Mills. Four Weeks ending Aug. 25th, 1888

Reduced to Kegs 25lbs. each

1888	Rifle	Shipping	Soda
Augt. 4	827		4008
11	332		4925
18	300		4757
25	1901		4876
	3360		18566

Total 21926 Kegs

Shipped to Mills during month					
	ending Augt. 25/1888		Saltpetre	1196	Bags
On hand	"	"	Saltpetre	2174	"
do	"	"	Soda	3269	"
Refined	"	"	Saltpetre	581	"
do	"	"	Soda	981	"

E&EO

New York Sept. 4th, 1888

The Hazard Powder Co.

pr. Tho. L. Lequin

Courtesy of Hagley Museum and Library

Mill Statement
Of Receipts & Expenditures for August 1888

Cash on hand August 1st.		$ 3,007.69
Received from New York Office		12,500.00
Received from other sources including rece'ts paid in		705.04
		$16,212.73
Contra		
Paid on Pay Roll	$7,195.55	
" for Teaming	865.76	
" " Freights	1,470.23	
" " Packages	473.86	
" " Mill Purchases	2,153.71	
" " Alder Wood	1,073.88	
" " Fire "	133.28	
" " Sundries	155.11	13,521.38
Cash on hand Sept. 1st		$ 2,691.35

Hazardville, Ct.
Sept. 11th, 1888

(signed) The Hazard Powder Co.
By W. B. Covil

Courtesy of Hagley Museum and Library

The Glaze Mill

Statement of Sales of Blasting and Sporting Powder made by The Hazard Powder Company during the calendar year 1888 on which percentages of increase or decrease as against the yearly average of 1882, 1883 and 1884 were based in reports made to the Depositary under the agreement of date Aug.3d, 1886, made between the 3 and 9 Cos. so called.

Blasting	236553 Kegs of 25lbs each
Sporting – equal	68925 " " " "

State, City & County] S. S.
of New York]

I, William S. Colvin, Treasurer

of The Hazard Powder Company do solemnly swear that the above

statement is true and correct to the best of my knowledge and belief.

W. S. Colvin

Sworn this 21st day of December, AD 1889

Before me J. A. Hillery, Notary Public NY Co

Office of

The Hazard Powder Company

Manufacturers of Gunpowder

Edward Prickett, Supt.

Mr. C. D. Prickett, Ast. Supt. Hazardville, Conn. July 20th, 1895

Mr. W. S. Colvin, President,

#44 Cedar Street

New York City

Dear Sir,

In compliance with yours of 17th, inst. I have made out the following memo, which I hope may be of service.

The Hazard Powder Co's plant was begun about the year 1835, and the names of the persons composing the Company were, Parks Loomis, Allen Loomis, Neland Loomis, A. G. Hazard, and A. A. Denslow, and the Company was called the "Loomises, Hazard & Co."

The property was transferred to A. G. Hazard and A. A. Denslow in 1843, and by them at once passed over to The Hazard Powder Co. a joint stock company under the same name.

In 1843 the capital stock was $80,000, or 3,200 shares $25 each and owned as follows,

Mr. A. G. Hazard,	860 shares
Mr. A. A. Denslow	860 shares
Mr. W. J. Cornell	400 "
Mr. P. C. Cornell	400 "
George L. Cornell	200 "
Mr. T. T. Tompkins	280 "
Mr. S. G. Cornell	200 "

The officers were	Mr. A. G. Hazard,	President
	Mr. W. J. Cornell	
	Mr. A. A. Denslow	Directors
	Mr. S. G. Cornell	

The Loomises were all left out in 1843.

When William and Henry Prickett came to the mills in May, 1836, the mills were running and making powder with a cylinder mill, that is the incorporation was done in barrels or cylinders with composition brass balls. The first wheel mill was built in 1836, one pair wheels. Two pair more were started soon after say in 1837 near the barn.

In 1843 the upper stone dam was built and two pair more wheels started in what we now call No. 9, so that there was 5 pair wheels running in the year 1844.

From 1850 to 1854 the works were considerably enlarged with the addition of 6 pair wheels, 4 pair to run via steam and 2 pair by water, making 11 pair wheels running. Later one pair wheels was put in what is now called No. 8 and in 1876 this was made two pair.

In 1871 and 1872 the upper steam works were built, with 4 pair wheels, later the two pair near the barn were moved up and attached to these, making 6 pair at the upper steam works but two less by water, and making the number of wheels at Hazardville to 17 pair, since 1876.

Of course the other departments of the plant were enlarged from time to time to meet the requirements of the increased number of wheel mills.

Scitico Mills.

The Scitico Mills were built in 1849 and started in the spring of 1850 under the name of "The Enfield Powder Co."

The promoters of this Company were, Paul Greeley and Sylvester Viets, two men formerly in the employ of The Hazard Powder Co., P. Greeley as book-keeper and S. Viets as foreman of the mills.

Stock-holders were, P. Greeley,
S. Viets,
John King,
Tuder Gowdy,
Eli Gowdy,
Wells Loomis.

About Jan. 1st, 1854 this plant was sold to The Hazard Powder Co., and the Enfield Powder Co. went out of existence.

There were two pair wheels there then and in 1855 one pair more was started, making three pair at those works and these added to the 17 pair in Hazardville makes the 20 pair now in running order.

Canton Mills.

I can't give the early history of the Canton Mills, but they were transferred in 1843 to A. G. Hazard and A. A. Denslow by the Loomises, Hazard & Co. same time as the Hazardville mills were. In the early history of the mills, a cylinder mill only was used, later a single pair of wheels. The mills were abandoned late in the sixties, say sixty seven and the place sold to Chamberlain in the interests of a R. R. Co., in 1874. Machinery of any value was brought to Hazardville.

East Hartford Mills.

These were built and owned by Hanmer & Forbes, when I do not know. The Hazard Powder Co. bought them not far from the year 1860, and not much was done there after about 1870 and in 1880 the plant was sold and the machinery moved to Hazardville. There was only two pair wheels at that place.

Freighting

Powder has been sold by peddlers, but the Company had none in their employ.

We had magazines on the bank of the Connecticut River in Enfield for powder and storehouse for saltpeter and brimstone. Teams took the saltpeter out to the mills and powder in to these magazines, about 4 miles. Flat boats were used from Enfield to Hartford, and transfers made there to and from boats and schooners. The schooners took the powder to Constable Point, N. J., which was the general distributing point before railroads carried powder.

About 1857 we began shipping via rail from Springfield, Mass., over the Boston and Albany road, but two or three years later, that great explosion of high explosives in Worcester drove us off.

We then went by team to Stafford Springs and used the Vermont Central until the Connecticut Central, (now a part of the N. Y. & N. E.) was built, and since then we have used the railroads exclusively for freighting powder.

We feel quite sure there was no change in ownership from the origin of the Company down to 1843, for we find deeds conveying property on the Connecticut River to the three Loomises, and A. G. Hazard and A. A. Denslow in 1836.

Mr. A. H. Bullock appears as a stock-holder in 1845 and A. E. Douglass in 1846.

Now from the foregoing you may select and make out what I hope may be satisfactory to pass over to Mr. F. G. Du Pont. If there is anything else you want I will do my best to obtain it.

Very truly,

(signed) Edw. Prickett, Supt.

Courtesy of Hagley Museum and Library

Office of

The Hazard Powder Company

Manufacturers of Gunpowder

Edward Prickett, Supt. *Hazardville, Conn.* **Sept. 26th, 1896**

C. D. Prickett, Asst. Supt.

Mr. Francis G. DuPont,

Wilmington, Delaware

Dear Sir:-

Yours 24th, inst. received and I note that there will be a little delay in forwarding plans for the Press building, and of course in the meantime I can do nothing about the foundation to the building, or ordering the timber for the frame of the building, but we shall not get out of a job; we have plenty to do.

I am aiming to get the two East mills and the East battery done first so that if necessary, we can run four pair wheels soon, and send the product elsewhere for manipulation, for our water is running low and the water mills make poor showing.

We find all four of the spindles bent and must be replaced and the broken gear overhead in the West mill that we thought would do with two new segments when you was here, will require four new segments or the whole.

The gear is old. It came from our Canton, Conn. Mill very many years ago, is light and two of the segments had been broken and banded before, and the teeth are about half worn out, and all things considered, we concluded to get all four new segments which means the whole wheel.

When here you asked me if the West mill had ever exploded with those on the East side of the Engine. I think I answered once during a strong West wind, but if I so answered, I was in error. The two East mills have gone together, but fire

was never carried over the Engine house and batteries from the West mill to the East mills nor from the East mills to the West.

Now to your enquiry about the storm of a week ago this morning. I have made inquiry of quite a number of level headed men so as to be sure, and have my own opinion confirmed for at such times when laboring under such excitement, we may be wrong, but the almost universal opinion is that although the deluge of rain continued for some minutes after the explosions yet the lightning ceased from the cloud hanging over the place.

Some think they saw the second bolt come down but I think that this was the flame from the Press going up following the explosion at the Corning mill although it is possible that a second bolt came down. Many saw the flame of the Press as it went off.

We shall push work on the East side of the Engine and Boiler houses, so as to be ready to run in a short time if need be, for out trade is fairly good. We have shipped over 8400 kegs this week and nearly 22,000 kegs within the past three weeks, and we may want these mills soon.

I am very glad to see the stuff moving off for I would rather others would burn it than have it burnt on our plant as it did one week ago this morning.

PRESSPLATES: Our Press Plates are 24 inches square, 2/10 inches thick. Will the new Press require the same?

Price of new Plates is 55 cents per pound, new which is quite a reduction from former prices. We are offered 25 cents per pound for the damaged Plates.

Respectfully, Your obedient servant,

Edw. Prickett, Supt.

Courtesy of Hagley Museum and Library

Explosion at the New Works Wheel Mill – September 19, 1896

Remnants of the Wheel Mill – the New Works – September 19, 1896

RECEPTION

TO

SUPERINTENDENT EDWARD PRICKETT

BY THE

EMPLOYEES OF THE

HAZARD POWDER COMPANY

Institute Hall, Hazardville, Ct., April

Twenty-sixth, Nineteen hundred two

Employees of the HAZARD POWDER CO., Employed at their Mills in Hazardville, CT, and Scitico, Ct., April 22nd, 1902, Edward Prickett, Superintendent.

Office Hands

C. D. Prickett, Asst. Supt.
H. W. Miner
Jos. Brown
A. E. Colvin
E. L. Prickett

Powder Hands

James Colburn
John Claney
James White
Phillius Carrignen
Jos. Trudeau, Jr.
Burdett Gowdy
John Luffram
Charles Vasseur
Barnard Turner
Chas. J. Ryan
F. E. Turner
Donald Campbell
George Knight
Jos. Pothier
Alfred Blunden
Jos. Trudeau
Ludger Roulx
Henry Lahay
Thos. Dowd
Geo. H. Blunden
Chas. N. Colburn
Matt. Cunningham
B. W. Colburn
George Rivet
Louis Myers
James McElroy
Jos. Lozeau
Wm. Reynolds
T. J. Kennedy
George Sherwood
Louis Stocker
Jos. Polardis
Jos. Fisher
Chas. Brown
Elmer Hale
Wm. Smith
John McAviney
Eli Robinson
Emil Kude
Louis Filossie
Thos. Lawlor
Edw. Barrett
George Reynolds
Jno. S. Partiss
C. H. Blunden
Chas. Gaskell
Wm. St. Petre

Scitico

H. A. Colvin
A. E. Leonard
E. B. Moeller
Francis Barrett
W. H. Taylor
J. P. Dooley
George Grenier
George C. Richards
Wm. Robb
Roy Colvin

Engineers

James C. Pierce
Geo. W. Green
John Hayward
Thos. Lanham
Ernest Hayward
James Morris
Thos. Hayward
Wm. McDougal

Refiners

T. F. D'Arsey
E. R. French
John Kearns
T. A. Kearns

Edw. Ryan
Tim. Sweeney
Alfred Haywar
Adolph LaBelle

Jacob Stocker
Peter Thome
Phillip Pothier

Colliers

John Cunningham

S. B. Humes

Painters

Wm. S. Harrison

R. M. Barrett

Mechanics

R. M. Stratton
M. M. Farnham
A. S. Buck
C. H. Thompson

Wm. Reid

G. A. Stratton
Enoch Turner
Frank Pease
H. G. Rosenberger

Coopers

Albert Miller

John Bernier

Out Door Hands

Geo. H. Payne
Geo. C. Whiton
Herbert Trudeau
Illrick Trudeau
Dollar Paro

Wm. Morris
Oscar Toutant
Delmore Hawthorn
Fred Partiss

Keg Makers

P. F. Houlihan
Pat. Quinn
Pat. McCarthy

George Walker
Arthur Gaskell
John E. Ryan

Labeler

George Bridge

This document survives as the only roster of employees, but reflects the black powder industry in decline, since there were reports of in excess of 300 employees during the Civil War. On the original program the front indicates the date of April 26 and inside the date is listed as April 22nd.

EDWARD PRICKETT,

Superintendent of Mills The Hazard Powder Co., 1896.
Appointed in 1859.

Reception *to* Supt. Edward Prickett *by the* Employees *of the*

Hazard Powder Company

Institute Hall, Hazardville, Ct., April Twenty-sixth, Nineteen hundred two

FAIRLAWN GUNPOWDER
F.F.G.
MANUFACTURED
BY THE
HAZARD POWDER COMPANY.

The following pages reflect discussions regarding the retirement of the President William S. Colvin and the financial arrangements entered into between The Hazard Powder Co. and said Mr. Colvin.

This 'Plan for the arrangement with Mr. W. S. Colvin was written by Mr. Colvin himself according to correspondence of August 18th which follows.

PLAN FOR ARRANGING WITH MR. W. S. COLVIN

Mr. W. S. Colvin to have the right to take for himself, three months of every year, it being understood that he is not to be expected to work regularly or steadily but to give duPont and Company such time as they want him to at the present rate, namely $11,000 per annum, (or $36.00 per day), for such time as he may give duPont & Company to have the right to call on him for themselves or any of their allied interests.

Courtesy of Hagley Museum and Library

PERSONAL

Wilmington, Del., August 18, 1902

Mr. W. S. Colvin, President
Hazard Powder Company,

New York City.

Dear Sir:--

I am in receipt of your suggestion of the 18th inst. as to compensation after retirement and have discussed this matter with one or two of our people, but do not think at this time I will take it up with our stock-holders, as I could not myself advocate the arrangement which you suggest. The salary you suggest, added to the amount which had been agreed upon to pay Mr. Weightman would make the charges of the Hazard Company such that with a few more expenses of the kind so over laden it that I am inclined to believe we would be "killing the goose that laid the golden egg"; and, when we add to that ten dollars a day for the time we would likely want you to work for perhaps the next two or three years, if not longer, would from a business stand point make the proposition one which I would not care to recommend.

I want you to feel that the stock-holders of the Hazard Powder Company are anxious to do what is fair and just, realizing as you do that you have been fully paid for your services for the past forty years, there is really nothing due. But the stock-holders of the Hazard Powder Company want to offer you an annuity that would enable you to feel that you would not be called upon to do any work, and, an annuity that would enable you to live as you have lived, but, it is not their thought I feel sure to make the annuity such that it would be a money-making arrangement.

If the minimum fixed upon in the suggestion which I left with you is too low I would be willing to take up with our people the proposition to call on you for as many days at $25.00 (I am getting between my $35.00 and your $10.00) per day that would be necessary to make your minimum up to say $3,600.00 per year and while this may seem to you very low, I am honest when I say in my judgement it is a very liberal proposition. Of course, you could go into any business you might take a fancy to, from which you could derive an income dependent upon the business venture, providing such business is not antagonistic to the business in which myself and associates are engaged or interested.

I am perfectly willing, Mr. Colvin, if you want me to submit your proposition to our Directors at our regular meeting and for you to attend this meeting. I am only one of the six or seven interested in this and I do not think my judgement in these matters (owing to the short time I have been in the business) is to be taken as against those of longer experience.

I have covered in this letter I think in a general way my position clearly and have spoken very frankly, and hope you will take the letter in the spirit in which it is written, and trust you will be equally frank with me.

Yours truly,

T. C. duPont, Prest,
E. I. du Pont de Nemours & Co.

W. S. Colvin, President
Geo. Weightman, Treasurer

John L. Lequin, Secretary
Edw. S. Lentilhon, Asst. Secy.

THE HAZARD POWDER COMPANY,
MANUFACTURERS OF GUNPOWDER
GENERAL EXECUTIVE OFFICES
44-46 & 48 Cedar St.

All grades of
Black Blasting & Sporting Powders
Also
Smokeless Powders for
Shot Guns, Rifles and Revolvers

New York Aug. 26th, 1902

Mr. T. C. du Pont, Prest.
Messrs. E. I. du Pont de Nemours & Co.,
Wilmington, Del.

Dear Sir:-

Failing in having an intended short talk with you this morning, I – in order to inform our clerks, who feel that they cannot be kept in suspense any longer but must be on the lookout for new positions – now write for the following information, or so much of it as you are ready to impart:

1. Have you decided when you will move this office to Wilmington; if so, what date?
2. How long thereafter have you planned to have present salaries continued, or to pay them in advance as a bonus?
3. On what terms would you arrange for temporary or permanent employment of any who would go to Wilmington of those whom you would desire.

There have been so many rumors, more or less direct or suggestive, yet nothing positive, that much unrest abounds.

Until recently, there was a general understanding that the removal would occur when the Equitable Building had been enlarged, the date for which was named about November or December next, therefore no immediate effort was made for change of residence or position. The date, if to be earlier, causes demoralization as you can imagine, hence this letter, with regrets for its necessity.

Very truly yours,
W. S. Colvin, Prest.

Courtesy of Hagley Museum and Library

Wilmington, Del., August 27, 1902

Mr. Wm. S. Colvin,

46 Cedar St., New York

Dear Sir:-

I have your letter of the 26th instant under personal cover, in which you say that the figure named by you was intended to cover the precedent which has already been established. This is what I want to get at. Won't you please give me more information on this line? As well as I have been able to ascertain, the biggest salary that has ever been paid to anybody who has been retired, is $3,600. If I am wrong I should like to be set right.

Yours truly,

T. C. du Pont, Prest.

Courtesy of Hagley Museum and Library

Wilmington, Del., August 27, 1902

Mr. W. S. Colvin,

46 Cedar St., N. Y.

Dear Sir:-

I have your letter of the 26th instant. I have stated to each and every man in the Hazard Office except one man who was on his vacation just what our position is and unless they have very poor memories they ought to know. If any of them are dying to get away, let them go. We will help, as I stated to them, each and every one of them to secure a position.

Referring to the second paragraph of your letter, "or to pay them in advance as a bonus?" I suppose this is a bonus for leaving us in the lurch after having paid them well for years past.

Yours truly,

T. C. du Pont, Prest.

Courtesy of Hagley Museum and Library

W. S. Colvin, President
Geo. Weightman, Treasurer

John L. Lequin, Secretary
Edw. S. Lentilhon, Asst. Secy.

THE HAZARD POWDER COMPANY,
MANUFACTURERS OF GUNPOWDER
GENERAL EXECUTIVE OFFICES
44-46 & 48 Cedar St.

All grades of
Black Blasting & Sporting Powders
Also
Smokeless Powders for
Shot Guns, Rifles and Revolvers

New York Aug. 29th, 1902

Mr. T. C. du Pont, Prest.
Messrs. E. I. du Pont de Nemours & Co.,
Wilmington, Del.

Dear Sir:-

I have this afternoon had a plain and pleasant talk with each of our office force, and find them generally disposed not to leave us in the lurch but to aid either at this office or at Wilmington, one or both, in balancing our books and straightening up matters generally.

I have reported the result of our interview, as I understand it, namely, as follows:

That the office business will likely be moved from New York to Wilmington by the 10th, to the 15th, of September, if not earlier, and that to those who remain and assist, present salaries will be continued for the month of September, although employed only part of the month.

That is they are wanted at Wilmington, their expenses down, while there, will be paid by us, and salary for the full month allowed them.

Further than that, I am compelled to refer to your interview with each one, from which they understand that you will give them a permanent position in the Wilmington office if they care to remain.

<u>Book-keepers</u>: Mr. A. L. Miller (Reed) refers particularly to his understanding with you of to-day, in that he will go to Wilmington permanently on the terms you promised him.

Mr. W. Guinand, Assistant, would be willing to go to Wilmington temporarily only; he would be quite willing to do that to help us out.

General Clerk: Mr. J. H. Prentice is disposed to do all in his power here at the office but will not leave New York.

Shipping Clerk: Mr. Dudley Prentice refers to his interview with you - of today, which was not completed; he is to write you. He is willing to help us out, either here or at Wilmington.

ditto. Assistant: Mr. G. B. June. Will go to Wilmington, same position and salary as at present.

Mailing Clerk: Mr. H. A. Fales goes with Messrs. Lequin and Lentilhon.

Stenographer: Mr. Chas. V. Brooks is willing to help us out either here or at Wilmington.

The atmosphere having cleared somewhat, there will be no necessity – for the present at least – of your sending anyone from Wilmington; we can balance the books and account here, and send them down to you in good shape.

Yours truly,
W. S. Colvin, Prest.

Courtesy of Hagley Museum and Library

My dear sir,

Your plain terms of today placed me in a position so much needed. After an interview with each one, when I made them understand that if disposed to leave us in the lurch a force would be sent up from Wilmington to complete books and accounts here. I discerned a change of heart and a willingness to take what they could get if they could not have what they wanted.

I am very glad of this, both for your sake and mine.

Sincerely yours,

W. S. Colvin

Mr. T. C. du Pont

29, Aug. 02

Courtesy of Hagley Museum and Library

Wilmington, Del., August 30, 1902.

Mr. Wm. S. Colvin

46 Cedar St., New York.

Dear Sir:--

I am in receipt of yours of the 29th and note contents thereof in regard to employees of your office.

I hope to be able to move the business from New York by the 10th of September. Will start on it earlier. I note what promises you have made to them.

Referring to the last paragraph on the first page of your letter – it is in line with what I said except I also stated that the only thing that would keep them in a position was their ability to hold it down.

Mr. A. L. Miller was to write me and I hope he will do this, setting forth in writing what my promises were. Please thank Mr. Guinand for his promise to come and help us out.

I have referred your letter to Mr. Dunham.

Yours truly, T. C. duPont, Prest.

Wilmington, Del., September 1, 1902

Mr. W. S. Colvin,

46 Cedar St., N.Y.

Dear Sir:--

I brought up the matter of our retirement from business at our last Directors' Meeting and they entirely approved of the arrangement talked over at the meeting between yourself, Col. H. A. duPont and the writer. This arrangement we will be pleased to have take effect whenever you care to have it go into effect, not later, however, than September 30th of the current year. We will try to arrange to move the auditing department over here just as soon as possible.

The arrangement first above mentioned is as follows:-
That you are to be retired from the active service of the Hazard Powder Co. on an annual salary from them of $4,000 per annum, payable in monthly installments. The Hazard Powder Co. or E. I. duPont deNemours & Co. to have the right to call on you after giving due and proper notice for services in the way of advice, consultation, to act as director and for such other information as you may be able to give to them. Not, however to be of such a nature as to require constant and steady employment from you. It being distinctly understood you are at least to have three or four months continuous holiday to yourself in which to make a trip if you like.

If this does not in every way correspond to your understanding of the arrangement, won't you kindly write me?

Yours truly,
T. C. duPont, Prest.
E. I. duPont deNemours & Co.

Courtesy of Hagley Museum and Library

Wilmington, Del., September 20, 1902

Mr. P. S. duPont, Treasurer,

The Hazard Powder Co.,

Building.

Dear Sir:-

Please put Mr. W. S. Colvin on the pension list of the Hazard Powder Co. at a salary of $4,000.00 per year, payable monthly.

Please put Mr. Geo. Weightman on the same list at a salary of $2,500.00 per year, payable monthly.

Yours truly,

T. C. duPont, Esq.

Courtesy of Hagley Museum and Library

W. S. Colvin, President
Geo. Weightman, Treasurer

John L. Lequin, Secretary
Edw. S. Lentilhon, Asst. Secy.

THE HAZARD POWDER COMPANY,
MANUFACTURERS OF GUNPOWDER
GENERAL EXECUTIVE OFFICES
44-46 & 48 Cedar St.

All grades of
Black Blasting & Sporting Powders
Also
Smokeless Powders for
Shot Guns, Rifles and Revolvers

New York September 25th, 1902

Mr. T. C. du Pont, Prest.
Wilmington, Delaware.

Dear Sir:-

Referring to our verbal of this morning, I hand you herewith, for use at such times as may suit your pleasure, my resignations as Director, President, Treasurer, Secretary, Executive Committee, Member, for one or more, in thirteen separate concerns; twenty one resignations in all. Possibly I have omitted some, but I cannot call any to mind at this writing. I enclose same in one envelope, duly backed in such a way so that you may check them off as you use them. I beg that you will notify me promptly when, and what ever you may do as to the Hazard Company, so that I will not extend my official signature beyond the time it should cover.

Permit me to comment on a few points: I am Secretary of the Eastern Dynamite Company as a matter of convenience with power to sign such checks and papers as are presented to me from 99 Cedar St. for that purpose.

Lake Superior Powder Company, under the present conditions I have the same signing powers as Executive Committee of this Company as I have of Secretary of the Eastern Dynamite Company.

Phoenix Powder Manufacturing Company. I have already explained that this concern should pay a salary to whoever shall care for its business in New York or thereabouts: particularly as Secretary, as there is considerable work attached to that office.

Anthracite Powder Company. I believe you understand

that Hazard Company have no stock in that concern and that its ownership is in its name because it seemed important that the DuPont and L. & R. should not be known. Whether the same necessity now exists, Mr. Weldy can tell you.

Advisory: You understand that the annual meeting takes place next week when a committee for the ensuing year is elected or rather that date is fixed, I believe you will ask to have changed. My resignation you can use or arrange with the meeting for a non-re-election.

Again asking that you notify me of any change as to the Hazard Company, I remain Yours truly, W. S. Colvin

New York, October 1, 1902

My dear Sir:-

In resigning my several official connections with "du Pont" interests, you understand, and I believe appreciate the fact that in deciding not to continue in active service with you, was because being in somewhat impaired health, and tired of the forty four years continued office work, I needed partial rest.

I want to thank you personally for the interest you have shown in my welfare, and your evident confidence in my integrity. I think as matters proceed and you learn more of the past, you will better appreciate the position in which circumstances have placed me. I am delighted that, while not officially employed, I shall be at your command.

Wishing you and yours a long life, the best of health and all happiness, believe me

Sincerely yours, *W. S. Colvin*

To. T. C. du Pont, Esq.

Courtesy of Hagley Museum and Library

Upper Falls

The Original Cylinder Mill at the Horseshoe Falls

Termination of the Company

WAIVER OF NOTICE OF SPECIAL MEETING OF THE BOARD OF DIRECTORS OF THE HAZARD POWDER COMPANY

We, the undersigned, being all the members of the Board of Directors of The Hazard Powder Company, a corporation organized under the laws of the State of Connecticut, do hereby waive notice of the time, place and purpose of a special meeting of the Board of Directors of The Hazard Powder Company and do fix the 29th day of OCTOBER, 1912, at 9:30 o'clock A.M. as the time, and the office of the company in Wilmington, Delaware, as the place of said meeting for the purpose of taking action in the matter of terminating the corporate existence of said company, winding up its affairs and distributing its assets among the stockholders, and we do hereby consent to the transaction of such business as may come before said meeting.

T. C. duPont
Pierre S. duPont
E. DuPont
F.I duPont or H.F.
Alexis I. DuPont

Dated: Oct. 29th 1912

Courtesy of Hagley Museum and Library

MINUTES OF SPECIAL MEETING OF
BOARD OF DIRECTORS OF
THE HAZARD POWDER COMPANY

HELD at Wilmington, Delaware, at 9;30 o'clock A. M. on the 29th day of October, 1912, pursuant to waiver of notice of said meeting signed by all the directors, at which meeting a quorum was present:

Messrs. P. S. du Pont
Eugene du Pont
Alexis I. du Pont

Mr. P. S. du Pont presided at the meeting with Mr. Alexis I. DuPont as secretary, present and acting.

The following preambles and resolutions were introduced and on motion, duly seconded, were unanimously adopted:

WHEREAS, the stockholders of this company have by an agreement in writing signed by them on the 28th day of October, 1912, consented to the dissolution of the corporate existence of the company and therein authorized and directed the Board of Directors of this Company to wind up the business of the company, collect its outstanding accounts, pay all of its debts and liabilities, and after payment thereof to divide the residue of its assets among its stockholders according to their respective holdings in said company, and to take all lawful action to terminate the existence of said company, and

WHEREAS, the total assets of this company consist of an outstanding account due and owing to this company by E. I. DuPont deNemours Powder company, a corporation organized and existing under the laws of the State of New Jersey, amounting to $1,002,668.82;

THEREFORE, BE IT RESOLVED that the officers of this company are hereby authorized, empowered and directed to proceed at once to collect said outstanding account of $1,002,668.82 from said E. I. DuPont deNemours Powder Company, and after payment of all the debts and liabilities of the company therefrom, to distribute the residue thereof among its stockholders;

And

BE IT FURTHER RESOLVED that the secretary of this company be and he is hereby authorized, empowered and directed to cause to be published in a newspaper published in Enfield, Connecticut, and having a circulation in Hazardville, Connecticut, where the principal office of this corporation is located, a notice to creditors, in accordance with the provisions of Section 30, Chapter 194, Public Acts of Connecticut, 1903, requesting all creditors to present their claims against this company to Mr.

Edward L. Prickett whose post office address is Hazardville, Connecticut, within four months from the 28th day of October, 1912, to-wit: on or before the 28th day of February, 1913, and stating that in accordance with the provisions of said act all claims not presented within the time limited will be forever barred; and

BE IT FURTHER RESOLVED that a majority of the directors of this company shall immediately after publication of said notice, as foresaid, make, sign and swear to and file in the office of the Secretary of State of the State of Connecticut, a certificate that stockholders agreement and consent that the corporate existence of this corporation shall be terminated has been duly given, and that all claims against said corporation may be sent to Mr. Edward L. Prickett, Hazardville, Connecticut; and that when the directors have completed their duties as aforesaid, a majority of the directors of this company shall make, sign and swear to and file in the office of the Secretary of State of the State of Connecticut, a further certificate stating that the directors have completed their duties as prescribed by Sections 30-34 of "An Act Concerning Corporations" (being Chapter 194, Public Acts of Connecticut 1903), in winding up the affairs of said corporation and have sold or collected all of its assets and have distributed the same among its stockholders according to their respective holdings in said company. There being no further business, the meeting adjourned.

Alexis du Pont, Secretary
Approved. P. S. du Pont, Chairman

Courtesy of Hagley Museum and Library

TO THE CREDITORS OF THE HAZARD POWDER COMPANY:

Notice is hereby given to the creditors of The Hazard Powder Company that the stockholders of The Hazard Powder Company have unanimously agreed by agreement dated the 28th day of October, 1912, to terminate the corporate existence of this corporation, and the creditors of this corporation are hereby warned in accordance with the provisions of Section 30, Chapter 194 of the Public Acts of 1903 of Connecticut, and in further accordance with the vote of the directors of this company passed at a meeting held on the 29th day of October, 1912, to present their claims against this corporation to Edward L.Prickett, its agent, at his office at Hazardville, Connecticut, within four months from the 28th day of October, 1912, to-wit: on or before the 28th day of February, 1913.

In accordance with the provisions of said act, all claims not presented within the time limited above will be forever barred.

Dated this 29th day of October, 1912.

Alexis I. DuPont

Courtesy of Hagley Museum and Library

Explosion at Powder Mill

The Final Blast

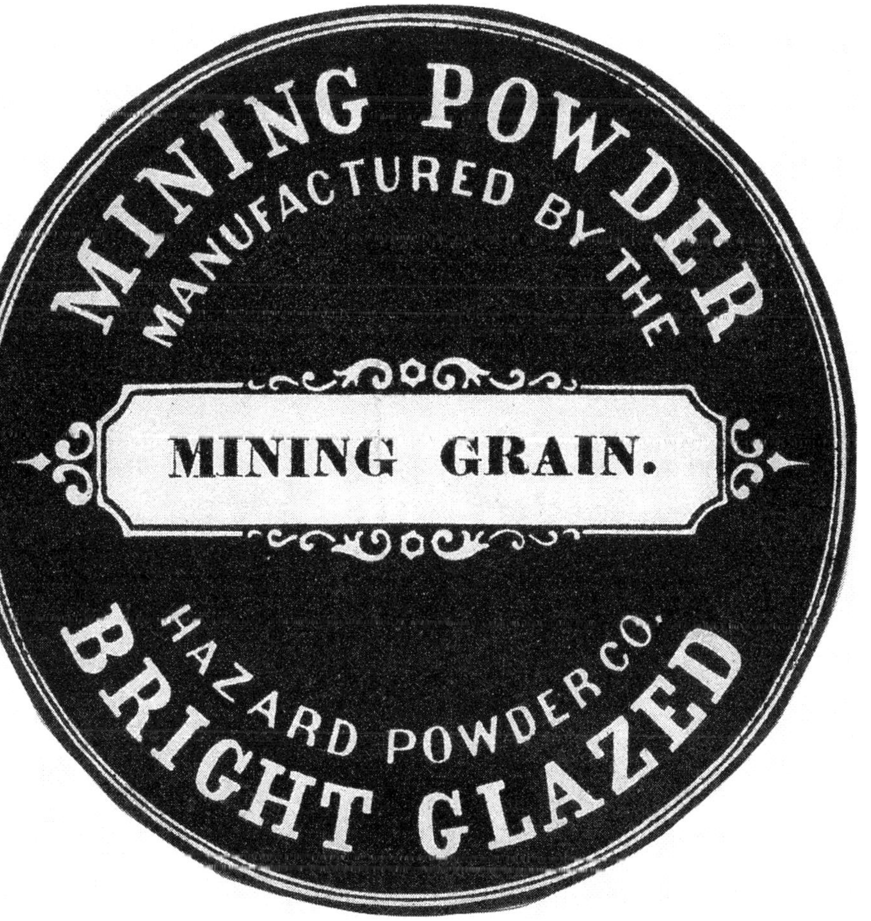

THIS EDITION CONTAINS TEN PAGES

The Thompsonville Press

ESTABLISHED 1880 — THOMPSONVILLE, CONN., THURSDAY, JANUARY 16, 1913

Four Hazardville Powder Mills Blown

Two Men Instantly Killed

Other Workmen are Injured But None Seriously

Worst Catastrophe in Hazardville's History---Almost Every Window in Hazardville Shattered---Explosion of 1500 pounds of Power, Heard and Jar felt as far as Springfield and Hartford---Hundreds Hasten to Scene---Victims Each Leave Large Families---Funerals Will Be Held Tomorrow.

About half past one on Tuesday afternoon people in this vicinity, and in fact for some miles around, were startled by a terrific explosion that rattled windows and jarred buildings to their very foundations, and it was

JACOB STOCKER

[illegible] Powder Explosion at Hazardville

[illegible] found.

CHARLES BLUNDEN

Second Victim of Powder Explosion at Hazardville

The cause of the explosion will, in all probability, never be known. Two men, Charles Blunden and Jacob Stocker, were at work in the press mill and to them alone could the

REV. DAVID LEWIS YALE RESIGNS PASTORATE

First Congregational Church to Lose Greatly Beloved Minister

Resignation Which Will Take Effect June 30th Will Close Nearly Five and One-Half Years of Most Successful Work by Pastor and People—Pastor Yale's Work Widely Known Throughout the Town and State.

The resignation of the Rev. David Lewis Yale as pastor of the First Congregational church, of Enfield, came as a complete surprise to his wide circle of friends and admirers in this town as well as to the church and congregation he has served so faithfully. It was read by Mr. Hatheway, at the annual meeting of the Ecclesiastical society last Thursday evening, after the society had voted to increase the pastor's salary. Mr. Hatheway, to whom the resignation had been handed, had not opened it before and supposed it had to do with another matter.

The letter of resignation follows:

"To the First Congregational Church and to the First Ecclesiastical Society, of Enfield, Conn.

"Dear Friends: Five years ago next month you called me to be your minister. Tonight I resign that sacred trust, to take effect June 30.

"The situation here five years ago called for forceful leadership in both pulpit and parish and that I have tried to render. To my efforts you have responded nobly. The results of our labors together are evident in our chapel, remodeled, beautified and increased in usefulness. They are also seen in our meeting house, thoroughly repaired, redecorated, and refurnished for Christian worship and service. These works have occupied much of our time. But our largest and our happiest work together has been in spiritual things. These results we cannot estimate for they are wrought in character, and fellowship, and faith, and hope, and love. God keeps these records for the life to come. He has given us encouragement in these labors, in the facts that our church membership, with 42 additions, is the largest we have reached during the last 12 years, while our Sunday school, where centers our best work for our young people, has improved in quality and has reached the largest enrollment for many years.

"My ministry in Enfield has been a happy experience. My reason for resigning is this: I believe that a minister should not be over long in a situation which compels the minister to be forceful in leadership, if he would do the task set before him.

"In behalf of Mrs. Yale and myself let me thank you for the many words and deeds of kindness by which you have brightened for us the five years now closing.

"Yours in Christian service,
"DAVID LEWIS YALE."

At the conclusion of the reading the society voted at once, and unanimously, not to accept the resignation, and not to enter record of it upon the minutes of the society, and appointed a committee to wait upon Mr. Yale and ask him to reconsider his action and if possible to withdraw the resignation.

After seriously considering the matter for two days, Mr. Yale sent the following letter to the society, which was read at the morning service Sunday:

"To the First Ecclesiastical Society of Enfield, Conn.

"Dear Friends: The committee appointed by you at your annual meeting last Thursday evening to confer with me regarding my resignation has brought to me your unanimous request that I reconsider and if possible withdraw it.

"Your heartiness in this action, your vote not to enter the announce-

The Cross Shows Location of Powder Mill in Which First Explosion Occurred. Nothing of Which Remains

committee last evening presented to Rev. Mr. Yale the following:

"Rev. David L. Yale,

"Dear Brother:

"Sincerely regretting the necessity which calls for this letter from the church you have served so successfully and acceptably for the past five years, it is still our privilege to testify to our appreciation of the work which you have accomplished, and to assure you that it will long be remembered by the old church in Enfield.

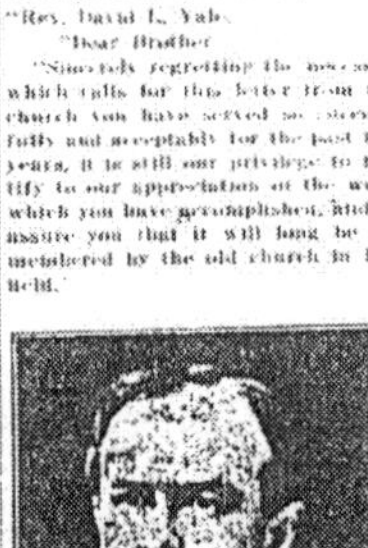

Rev. David Lewis Yale

"We highly appreciate the help we have received from you in remodeling and beautifying our church edifice and its furnishings, and we shall miss your skillful and artistic advice in all such matters.

"A large part of a pastor's usefulness lies in his close touch and interest in the home and social life of his people. We shall miss you in this.

"We shall miss your kind and cheery greetings, your constant care and sympathy in our times of sickness and sorrow, and your hearty entering into the joys and festivities of our lives.

"We shall miss your interpretations of the teachings of the Bible, and your instruction from the sacred desk, which has ever been an inspiration to higher Christian living.

"The interest manifested by both Mrs. Yale and yourself in the life and activities of the church and community form a bond strong and enduring, and will be remembered with gratitude and love, and we ask the continued interest in your thought and your prayers.

ASSESORS SUBMIT GRAND LIST

Exceeds Grand List of Last Year By $269,145

Heaviest Taxpayers the Hartford Carpet Corporation and Westfield Plate Company—A List of Those Who Bear the Greatest Burden of Enfield's Expenses.

The assessors have finished their work and filed their lists with the town clerk. Taxes have been increased on many property holders as indicated by the postal cards received Monday by property holders.

Horace K. Brainard is the highest individual taxpayer in Thompsonville, Cornelius J. Sullivan of New Haven, a former resident, coming next. The Hartford Carpet corporation's assessment is the largest, with $2,257,650, to pay a tax on. In Hazardville David Gordon is the largest individual taxpayer.

The following is a summary of the new grand list as compared with that of the previous year:

	1911	1912
Houses	$2,539,260	$2,682,000
Lands	579,277	602,807
Stores	327,075	361,175
Mills and manufactories	830,825	931,025
Horses	17,320	17,265
Cows	23,075	22,100
3 years old	2,115	[illegible]
2 years old	2,800	2,320
1 year old	1,190	1,360
Poultry, sheep and swine	2,910	2,015
Automobiles	25,600	37,300
Carriages	7,961	7,575
Watches and jewelry	3,125	3,450
Musical instruments	9,300	8,950
Average amount of goods	$13,722	$56,600
Investments in mechanical and manufacturing operations	913,800	913,950
Money at interest	15,385	9,350
Money on hand over $100	1,671	7,883
All other property	69,530	89,010
10 per cent. added	32,018	16,525
Total	$6,396,695	$6,665,810

The following is a list of the heaviest taxpayers and the amount on which they will be obliged to pay:

Hartford Carpet corporation, $2,257,650; Hazard Powder company, $128,125; A. D. Bridge Sons, incorporated, $161,850; Thompsonville Water company, $79,473; Upson-Martin Carpet company, $70,075; Gordon Bros., $78,100; Westfield Plate company, $55,600; David Gordon, $62,150; George H. Bushnell Press company, $15,800; Horace K. Brainard, $15,800; North family of Shakers, $18,100; Church family of Shakers, $58,315; South family of Shakers, $18,500; Hattie Lord Burbank, $15,100; Mr. and Mrs. Mau-

(Continued on Page Four.)

School Inves[...] Will [...]

FULL TEXT GIVEN [...]

Recommends New [...] ent Propositions [...] Is An Exhaustiv[...] Needs---Should [...]

The following [...] will be present[...] school committ[...] at the special [...] held Saturday [...] week.

To the Citizens [...] Town of En[...]

At the annual [...] Oct. 8, 1912, [...] tion was present[...]

"Resolved, Th[...] five, consisting of [...] men and two [...] town be appoint[...] matter of [...] mendations in Th[...] port back at [...] ing, with recomm[...]

The resolution [...] Messrs. James [...] Mulligan were [...] the selectmen [...]

The committee [...] meetings and ma[...] its to not only [...] Thompsonville [...] of the schools in Enfi[...] port herewith [...] embodies condit[...] and changes wh[...] the committee, [...]

The school at Hazar[...] in need of a [...] ing. Stoves now are [...] rooms of the b[...] same and it is [...] new system of [...] before Sept. 1, [...]

The committee would [...] ommend installi[...] the Hazardvill[...] for the use [...] but, owing to [...] sewerage syste[...] practical at pres[...] however, recomm[...] drinking fountai[...] take the place of [...] now in use.

Since sanitary [...] Enfield street bui[...] proved, the exit [...] such that the [...] school room on [...] obliged to go out of the [...] recenter at anoth[...] get at the toilet, wh[...] basement. This means [...] dren are at consid[...] weather, and it is [...] these conditions [...] the summer vacation.

There are abou[...] 1,28[...] tending the public school[...] of Enfield between the [...] 16 years and there ar[...] children, including th[...] over 16 years and those [...] town, making the tota[...]

From the Thompsonville Press, January 16th, 1913

Four Hazardville Powder Mills Blown to Pieces!

Two Men Instantly Killed

Worse Catastrophe in Hazardville's History --- Almost Every Window in Hazardville Shattered --- Explosion of 1500 pounds of Powder, Heard and Jar felt as far as Springfield and Hartford --- Hundreds Hasten to Scene --- Victims Each Leave Large Families --- Funerals Will Be Held Tomorrow.

About half past one on Tuesday afternoon people in this vicinity, and in fact for some miles around, were startled by a terrific explosion that rattled windows and jarred buildings to their very foundations, and it was soon ascertained that beautiful "Powder Hollow" had once more become the scene of a terrific explosion of powder. In the space of a very few minutes four mills of the Hercules Powder Co. were demolished, one, that in which the first explosion occurred, so completely that no vestige of it remains. The sides were blown entirely out of the other buildings, the wheel mills, but the end-posts, some beams and charred timbers still stand to mark the scene of the disaster. Of the two men who were hurled into eternity only small bits of flesh and bone clinging to charred fragments of clothing have been found.

The cause of the explosion will, in all probability, never be known. Two men, Charles Blunden and Jacob Stocker, were at work in the press mill, and to them alone could the origin of the fatal spark that ignited the powder have been known. The fact that both men were old-time powder workers, for many years in the employ of the company, and accounted among their most careful and trustworthy men, Blunden, in fact, being considered a specialist – convinces those who knew them that carelessness had no part in causing the explosion.

Jacob Stocker was a farmer as well as a powder maker, and owned a small farm on the road to Scitico and was employed as spare hand at the mills. He was born in Austria, but came to this country about 13 years ago. In his native country he was a member of the police department. He was married Aug. 24, 1903, and besides his widow he leaves three daughters – Hortense, aged 9 years, Emma 7 ½ and Louisa 6 years.

Charles Blunden was born in England, but came to this country in boyhood and was taken into the employ of the Hazard Powder company. The Blundens had been living in Wayne, N.J. for the past eight years, where Mr. Blunden was employed in the powder mills and had returned to Hazardville but two months ago. The fact that this was his first day's work in the press mill since his return and the fact that he was substituting for another man, render his a peculiarly sad case. He was working for Joseph Macnamara of Hazardville. Mr. Macnamara has a sister, Mrs. Joseph Garrow living in Thompsonville, who upon learning of the horrible accident, supposed, of course, that her brother was one of the victims. Mr. Blunden leaves a wife and three children of his own – Harriet, aged 9, Albert 6, and Harry 2: also three of his brother's children – Fred, aged 20, Edith 18, and Charles 15 – Mrs. Blunden's first husband having been a brother of the deceased. Alfred Blunden, the first husband, lost his life 14 years ago in an explosion of like nature in the same mill.

There are many conjectures, of course, as to the cause of the explosion, but they must ever remain mere conjectures. Some small particle of undetected metal in the powder mixture; some slight friction of the machinery – who can tell what? The fact that the high-pressure valve, which was found by an employe after the accident, was open an inch and a half, proves that it was not a high-pressure accident.

The scene of the accident defies description. Branches and limbs of trees, huge pieces of metal and charred timbers lay scattered all about, while parts of heavy beams were carried with terrific force across the Hollow and imbedded in the hard ground of the banks that form its enclosure.

In the village of Hazardville few buildings escaped some mark of the devastating power that was let loose. Windows were shattered in all the churches and many of the stores and dwellings, and scarcely a window was left in the hotel. The schoolhouse was badly shaken and flying glass made it dangerous for the inmates and the pupils were dismissed. The detonation was heard and the shock felt in Hartford and in Manchester, while in Rockville windows were shattered and bottles thrown from the shelves in a drug store.

A number of workmen were about the premises and some were injured, but only one seriously, and it is believed now that his injuries will not prove fatal. Joseph Trudean was driving a load of powder about an eighth of a mile from the place when he was thrown from the wagon, either by the force of the explosion or by the sudden starting of the horses through fright, and was badly cut and bruised and was at first supposed to be injured internally. He is an elderly man and has been nearly 50 years in the powder making business. Ulrie Trudeau, at work at the water mill, was thrown down and slightly bruised and cut. Robert Miller and Fred Chevalier were at work in the glaze mill and at the first explosion ran for their lives, fearing the igniting of the great store of powder under compression in their building. Dollar Paron, a carter, was on the grounds also but escaped injury. William Abel, the powder foreman, had a marvelous escape.

When the press mill exploded he stood at the door of the first wheelhouse, which was the second building to go. He was thrown to the ground by the force of the explosion, but regaining his feet he gained a point of safety just in time to escape the fate of

Blunden and Stocker, as three seconds after the first mill went came the second explosion followed closely by the third and fourth.

Charles Ryan and Joseph Fisher of Thompsonville, were wheelmen and they too escaped just in time. The smaller mills have been wrecked many times, but this is the first explosion in a press mill during work hours since 1875. About thirteen years ago a man named Bowers was blown into the rafters of one of the wheel mills and his eyes blown out and many bones broken. His death occurred the following day. September 19, 1896, lightning struck the plant and destroyed the press mill and caused fires, but no lives were lost as it occurred after six o'clock, when workmen were not about. By many who are conversant with the history of the mills this is considered the worst blow-up they have experienced.

Funerals of Blunden and Stocker.

Although only portions of the bodies of Charles Blunden and Jacob Stocker have been recovered, yet some pieces sufficiently large to afford identification have been fathered together by workmen who finished their search today. The funeral of Mr. Stocker will be held tomorrow morning at 9:00 o'clock in St. Bernard's church. Rev. Fr. Edward A. Flannery officiating, and that of Mr. Blunden will be held from St. Mary's Episcopal church at 1:30 tomorrow afternoon and Rev. D. Russ Judd, the rector, will officiate. Mr. Blunden leaves his widow and six children and Mr. Stocker leaves three children besides his widow.

Card of Thanks

We desire to extend, through the columns of The Press, our sincere thanks to the friends and neighbors who offered us consolation and the ready hand of sympathy in the hour of our sad bereavement in the sudden death of our beloved husband and father, Jacob Stocker, and especially do we wish to say a word of appreciation to the officials and employes of the Hercules Powder Co. for their sympathy and for their beautiful floral offerings at the funeral service.

Mrs. Jacob Stocker and Family

Card of Thanks

To the Hercules Powder Company and their employes, and to all our kind friends and neighbors who sent floral tributes, and in so many other ways manifested their sympathy for us in this our time of most grievous affliction, we desire to extend, through the columns of The Press our sincere thanks and sense of appreciation of their many kindnesses.

Mrs. Charles Blunden and Family

From the Thompsonville Press, January 23rd, 1913

Officials of the powder mills will give their decision soon as to whether they will rebuild the local plant destroyed in the recent explosion. About 1,000 persons from out of town came to view the ruins of the mills Sunday. Many interesting points about the explosion have developed in cleaning up the debris. A part of the press weighing between three and four tons, was found 75 feet from its original location. The brass rolls, which were four feet long and eight inches through, have not as yet been found. The barricades consisting of oak timbers standing 30 feet in height and driven eight feet into the ground, were broken into kindling wood. A piece of gearing weighing nine pounds dropped within eight feet of Mr. Payne while he was standing in his tobacco lot, at least 1,500 feet from the scene of the explosion. Mr. Campbell has the gearing on exhibition in his barber shop window.

Ruth Blunden
Photo by Edward C. Allen, II

Ruth Blunden

The women of Hazardville were pioneers, many of whom came by ship from a foreign country, following in their husbands footsteps and settling in a place with austere living conditions and dangerous work. At the first sound of an explosion in the mills of the Hazard Powder Company, the women of Hazardville would congregate in the center of the village hoping that their husband would be a survivor.

Ruth Blunden was one of those women. Ruth was first married to Alfred Blunden. Alf worked in Powder Hollow for Col. Augustus G. Hazard. Ruth, like so many others, suffered the loss of her husband at an early age and while pregnant with his child when one of the mills blew up.

Alf's brother Charles Blunden applied for his brother's job. In those days jobs were scarce and employment was limited. He also married his brother's widow, Ruth.

On January 14, 1913, Ruth heard the explosion in Powder Hollow and said, "I knew my Charlie was gone!" Indeed, Charles had not survived the last explosion of the Hazard Powder Company and Ruth became a widow of the industry for a second time.

In an interview with Mrs. Blunden, Ed Allen asked her what would happen to a widow of the powder company; "Would they provide a pension for you"?

"Oh no", Ruth answered. "They'd give us a job!"

"Well, what could a woman do for the powder company?" she was asked.

"We'd paste labels" was her answer.

'Would you be paid by the piece?"

"No", Ruth said, "we'd be paid by the day".

"And how many labels could you paste in a day"?

She answered, "If I tried real hard I could paste a thousand".

Ruth Blunden simulating pasting a label on a powder keg.
Ruth was 95 years old at the time of this interview.

Photos by Edward C. Allen, II

List of Casualties

PERSONS KILLED BY EXPLOSIONS AT HAZARDVILLE AND SCITICO, CONN.

Date	Names	Buildings
March 11, 1839	Hiram Pierce and Robert McChesney	Press
Fall – 1839	N. Dunn and Wm. Leanard	Graining Mill
Aug. 25, 1842	Geo. Prickett	Rolling Mill #10
June 10, 1944	Wm. Luce	Graining Mill
Oct. 16, 1844	Wm. Prickett) N. McClester) H. Dutfee)	Rolling Mill #10
March 31, 1845	Name lost. Man came from Schaghticoke Mills, N.Y	Graining Mill
July 9, 1845	Wm. Hall	Graining Mill
March 26, 1846	Dan Bert	Rolling Mill #9
Aug. – 1849	Jas. Bisker	Clothes caught fire noon hour
April 10, 1851	Jas. McKenzie) Jas. Pierce)	Rolling Mill #10
May 20, 1853	Wm. Murry	Rolling Mill, Scitico
June 9, 1853	Walter Luce	Graining Mill, Scitico
April 6, 1855	Horace Hazard	Killed at Mortar, proving powder
Nov. 28, 1855	James Murphy	With powder wagon on plant
July 12, 1856	Chas. Elcock) Wm. Lowrie) A.Whitehead)	Lower Engine, Rolling Mills
Oct. 27, 1857	John Burns	Rolling Mills, Scitico
Sept. 13, 1858	J. Garache, Supt.) Wm. Colvin, Foreman) Stephen Pace) Thomas Ball)	Mixing Mill, Experimenting
Oct. 19, 1859	Michall Burns	Rolling Mill #9
July 23, 1862	Arthur Beach) Jas. Beach) H. Clark) Pat Carney L.. Monseau) Edw. Gramond) Jas. Neelands) Pat Fallows) Miss. Smith)	Explosion started at Press. Two Presses, one Graining mill & two Rolling Mills, one following the other in quick . succession .
July 5, 1864	Jno. Howarth) Jno. Hazzerty)	Graining Mill
Dec. 22, 1865	Wm. Fahey	Mixing Mill

May 7, 1866	Jno. Koshan	
	Wm. Agnew	
	Edw. Burke	
	Pat Bailey	Press
Dec. – 1867	Peter Havey	Mixer
June 30, 1868	Frank Richardson	Rolling Mill, Scitico
Feb. 11, 1869	Jos. Yates	
	Thos. Turner	Rolling Mill #9
Feb. 11, 1870	Pat Turneran	Rolling Mill #9
May 30, 1870	Peter Ryan	Rolling Mill #10
May 25, 1871	G. Taylor	Graining Mill
	"Poney" Agustine Racine Loseau	Taylor was with horse and cart.
May 19, 1875	Moses Bolac	
	Jno. Lamers	
	Geo. Richards	Press
		First explosion remembered by E. L. Prickett – 5 ½ years old.
Aug. 14, 1883	Jas. Leanard	Graining Mill
March 16, 1886	Chas. Luce	Graining Mill
June 29, 1892	Dennis Bailey	Graining Mill
Aug. 9, 1894	Geo. Raiche	Graining Mill
June 18, 1897	Mat. Pierce, Foreman	
	B. Stratton, Machinist	
	Richard May	Graining Mill
May 27, 1898	Alf. Blunden	Graining Mill
Feb. 26, 1901	Eugene Bowen	Rolling Mill #9
Jan. 14, 1913	Jacob Stocker	
	Chas. Blunden	Upper Sporting Press

Thompsonville Press, January 23, 1913

HERCULES POWDER CO. NOT TO REBUILD

Men Will Be Placed at Work in Its Other Mills if They so Desire.

Superintendent E. L. Prickett of the Hazardville powder works informed the employes yesterday that the plant would be closed for an indefinite period. The employes will be given employment elsewhere in the various plants of the Hercules company as opportunity opens, and if they desire.

Twenty-seven men are out of employment as a result of the accident. The company will pay these men until the end of the present month, and the bonus agreement which the company makes with its employes will continue uninterrupted if the company rebuilds later and they return to its employ or if they are placed at work elsewhere in the mills of the company.

David Jellison looking through the remnants of a culvert in Powder Hollow

MAKING BLACK POWDER
AND
A GLOSSARY OF POWDER MAKING TERMS

Making Black Powder

Proportions - The proportions of the three ingredients vary, but a frequently used formula calls for three-quarters saltpeter, one-eighth sulphur, and one-eighth charcoal. The ingredients when refined are ready for mixing. If the sulphur and charcoal are not in powder form, they may be pulverized together in rotating barrels by the tumbling action of zinc balls, or under heavy pulverizing wheels. All ingredients are bolted or sifted to make certain they are of the proper degree of fineness.

Stamping Mill or Pounding Mill - The three ingredients are first mixed together according to formula with a spatula with water. Lumps are removed by use of a sieve and a roller. The mixture is then placed in a battery of mortars, water is added and the pestles are put in motion. The stamping continues for about 1 hour.

Mixing Mill

Rolling Mill or Wheel Mill- Instead of the stamping mill process the materials may be mixed in rolling mills, also called grinding or incorporating mills. These mills consisted of two vertical rolling mill-stones or cast-iron wheels set upon a bed-plate encircled by a trough. The two vertical wheels turn on the bed-plate, mixing or incorporating the powder ingredients by their great weight and abrasive action. The amount of material incorporated at one time was about 300 pounds and the running time was from one to three hours for every fifty pounds of composition, depending upon the type of powder being made.

Upper Double Wheel Mill

Pressing - As it comes from the rolling mill the powder is a mealy, moist "powdercake." To get rid of moisture and to increase its density it is taken to a press mill where it is placed in a screw or hydraulic press and subjected to heavy pressure. When taken out it has the appearance and feel of a piece of slate and is referred to as "presscake."

Press Mill

Granulating - This "presscake" is then broken up into small chips by hand mallets or by being run through a series of zinc rollers and then passed through a number of sieves or screens to obtain the right size of grain. The powder dust and undersized grains are returned to the press mill and added to the next batch of powder.

Crackers or Corning Mill

Drying - The powder still contains too much moisture so it is dried by the sun on outdoor tables or on racks in a heated building known as a dry house. To obtain thorough drying it is stirred or tuned over periodically.

Dry House

Glazing - Sporting and other high quality powders were further dried by being rotated in a glazing barrel for a number of hours. The tumbling action of the powder created heat, rounded off the grains, and imparted a luster to the finished powder. A later innovation was the addition of a small quantity of graphite to the powder in the glazing barrel. This coated the grains, making them more resistant to dampness, and enhancing the glossy sheen of the powder.

Glazing Mill

Packing - The final step consists of running the powder through a reel sieve to get rid of powder dust created in the glazing operation. It is then packed in canisters, kegs, or barrels, labeled, and stored in a magazine.

Packing House

Glossary of Powder Making Terms

Saltpeter - Potassium nitrate created by the decomposition of animal and vegetable substances, usually in warm, moist regions. Found principally in India but also in limestone caves of Kentucky, Virginia, Tennessee and Georgia. Refined saltpeter constituted about 75 per cent of black powder. Sodium nitrate, or Chilean saltpeter, was used after the mid-nineteenth century for making blasting powder.

Sulphur - An element usually found in volcanic areas such as Italy and Sicily. Refined sulphur constituted about 12 ½ per cent of black powder.

Charcoal - Branches of light woods such as willow, poplar or alder charred by the burning out of gaseous and liquid materials. Charcoal constituted about 12 ½ per cent of black powder

Refining - Purifying raw materials by boiling, followed by successive washings, evaporation and crystallization by which means foreign substances were removed. Sublimation is the term sometimes used in the refining of sulphur.

Pulverizing - Grinding the refined raw materials between millstones or beneath rolling wheels or rotated in barrels containing hard discs to obtain proper degree of fineness.

Bolting - Passing pulverized ingredients through screens of fine mesh to sift out extraneous materials and to assure proper fineness of grains.

Composition - A preliminary mixing of prepared saltpeter, charcoal and sulphur according to formula.

Stamping, Rolling, Incorporating, Grinding - Pounding or incorporating the ingredients into a homogeneous mixture by batteries of mortars and pestles or by rolling wheels. Water was added to prevent the mixture from heating and becoming explosive.

Powder Cake - Moist, "caky" powder removed from mortars of stamp mill or tub of rolling mill after incorporation.

Pressing - Subjecting powder cake to pressure exerted by screw press or hydraulic press to further blend ingredients, give density to the powder, and to eliminate moisture.

Graining, Granulating, Corning - Synonymous terms for screening powder into grains of various sizes by cutting powder cake into chips and passing them through sieves of varying gauges.

Drying - Exposing powder to heat of sun on outdoor tables or to artificial heat in a building known as a dry house.

Glazing, Polishing - Rotating powder in barrels containing graphite whereby grains are rounded and further dried through friction and take on a moisture-resistant coating. Powder subjected to this glossy treatment was known as glazed powder; powder not treated in this manner was known as rough powder.

Dusting - Final sifting of the finished powder to remove powder dust and undersized grains which were returned to stamp or rolling mill for reincorporation.

Packing - Weighing and packing powder into canisters, kegs and barrels of rifle powder, progressing from F, a coarse or large grain, to FFF, a very fine or small grain powder. The subscript "g" (viz., FFg) indicated it was glazed powder; the subscript "r"(viz., FFr) meant it was rough or unglazed powder.

Magazine - Building in which powder was stored awaiting shipment.

Remnants of the Man Killer
Photo by Edward C. Allen, II

Colonel Hazard's Last Will & Testament

The following document is a copy of Col. Hazard's last will and testament. It has been copied verbatim with misspelled words etc., just as he wrote it.

AUGUSTUS G. HAZARD'S WILL

I, Augustus G. Hazard, of Enfield in the State of Connecticut being of sound and disposing mind and memory, do make and publish this my last will and testament.

I hereby constitute and appoint Alexander H. Bullock of Worcester, State of Massachusetts, my wife Saloma G. Hazard, and my daughter Fanny Hazard and the survivors or survivor of them to be executors of this my will and in the event of the decease of said Alexander H. Bullock either before or after my decease, I hereby constitute and appoint my daughter Elvira, wife of said Alexander H. to be his substitute, and in the event of the decease of my wife Saloma G. whether before or after my decease I hereby constitute and appoint my grandson Augustus George Bullock to be her substitute, either or both of said substitutes to act with the same power and discretion as the origional appointees.

It is my will, and I so direct that my said wife Saloma G. Hazard shall have the right during her life time to remain in possession and enjoyment of my homestead including the land belonging thereto and conveyed to me by A. K. Parsons, the mansion house and other buildings thereon, all my furniture, library, horses, cattle, carriages and other articles of personal property in use upon said homestead, subject to the right of my daughters Georgeanna and Fanny or either of them, while unmarried, to enjoy jointly with my said wife, and during the life time of my said wife, said use and occupation, the ------- of household expenses and support to be agreed by and between them selves, after the payment of my just debts and charges. I dispse of my estate as folows;

I give, bequeath and devise to my daughter, Fanny, the said homestead, including the said land, buildings and personal property therein before mentioned, subject to the aforesaid right of my wife and of my daughter Georgeanna while she remains unmarried, and only during the life time of my said wife, to occupy and enjoy the same.

I give and bequeath to my said wife, Saloma, G. Hazard one hundred shares of the capital stock of the bank of America, New York, fifty shares of the capital stock of the Farmers and Mechanics Bank (National) of Hartford, one hundred shares of the capital stock of the Hartford and New Haven Railroad Company, one hundred and thirty five shares of the capital stock of the Western Railroad Corporation, and thirty shares of the capital stock of the Hampden National Bank, Westfield.

I give and bequeath to my grandson Edward B. Sexton one hundred shares of the capital stock of the New York and New Haven Railroad company, fifty five shares of the capital stock of the Boston and Worcester Railroad company and ninety six shares of the capital stock of the Philadelphia, Wilmington, and Baltimore Railroad Company.

I give and bequeath to my daughter Georgianna seventy shares of the capital stock of the Hazard Powder Company.

I give and bequeath to my said executors of the following shares of capital stock (to wit) two hundred and fifty four shares of the Hazard Powder Company, one hundred shares of the Hartford and New Haven Railroad Company, one hundred shares of the Boston and Worcester Railroad company, one hundred shares of the New York and New Haven railroad company, and one hundred and fifty two shares of the Philadelphia, Wilmington and Baltimore Railroad Company, the same to be held in trust for the benefit of my said daughter Georgianne during her lifetime, the income thereof to be paid to her as received by said executors, and after her decease the said stocks to be dispsed of as folows;

if she shall marry and have children, then, upon her decease, to each of her children here after born, and living at the time of her decease, there shall be madeover and tranfered a proportionate amount of all of said stocks of the value of twenty five thousand dollars if the number of such new children shall not excede three, but if they shall excede three in number, then the whole of said stocks shall be divided among them equally at her decease and whatever shall remain of said stocks after fulfilling these provisions for her future children, whether it may be the whole or a part thereof, shall upon the decease of said Georgianne be equally divided between my daughters Elvira H. and Fanny and their heirs by right of representation.

I give and bequeath to my said executors the following shares of capital stock, (to wit) three hundred and twenty four shares of the Hazard Powder Company, one hundred shares of the Hartford and New Haven Railroad company, one hundred shares of the Boston and Worcester Railroad Company, one hundred shares of the New York and New Haven Railroad company, one hundred and fifty two shares of the Philadelphia, Wilmington and Baltimore Railroad company, the same to be held for the benifit proportionately of my grandchildren Anna Mumford Dakin, Arthur Hazard Dakin, Ellie Dakin, and Emily Dakin, children of my deceased daughter Emily Hazard, the income to be applied in the discretion of said executors, for the support and education of said children, on any part, if practicable, to be accumulating with the principal, and as said children or either of them shall attain the age of twenty one years, or upon the marriage of either before attaining that age, then one fourth part of said stocks shall be transfered and made over to such child, and if either of said children shall decease before attaining the age of twenty one years the portion to which such child was entitled whether income or principal, shall be equally divided among the remaining children.

The foregoing mention of the shares of the capital stock of the Boston and Worcester Railroad Company and of the Western Railroad Corporation shall be understood to apply, in same manner, and with the same meaning as to the number of shares, given and bequeathed, to the new issue of shares under whatever name the certificates shall be made when these two corporations shall have been consolidated under the laws of Massachusetts.

I give and bequeath to my grand son Augustus George Bullock, my namesake, the sum of twenty thousand dollars,

I give and bequeath to the Right Reverend the Bishop of the Protestant Episcopal Church for the

Diocese of Connecticut and to the gentlemen associated with him as a Trustees of Donations and Bequests for Church purposes, the sum of five thousand dollars, to be held for the benifit of the Protestant Episcopal Church organized at Hazardville in Connecticut and called St. Mary's, the same to be kept as a fund of which the income only shall be applied for the support of public worship, upon condition that out of the income thereof the Church property of St. Mary's shall be kept at all times insured in good and reasonible company or companies for not less than the sum of six thousand dollars, and that the Church property shall be kept in good repair, and that the residue of said income shall be applied to the maintainance of public worship in the same.

I give and bequeath to my said executors the sum of two thousand five hundred dollars, with instructions that thing (note: should read they) shall within one year after my decease pay the same to properly authorized committee or persons for the support of public worship in the Methodist Episcopal Church at Hazardville, upon the condition that the said sum shall be kept whole as a fund of which the income only shall ever be expended in maintaining said public worship, my executors are hereby instructed to require proper guaranty for the fulfillment of this condition.

I give and bequeath the sum of two thousand five hundred dollars to the first Congregational Church and society in Enfield, meaning that in which myself and family have attended public worship during our residence in Enfield, the same to be kept as a fund well and securely invested, of which the income only shall ever be applied to the support of public worship, I instruct my executors to pay said sum over to t he properly constituted officer or officers of said society upon receving satisfactory guaranty for the literal application of the bequest

All the residue of my estate I give and bequeath to my two daughters, Alvira Hazard, wife of Alexander H. Bullock and Fanny Hazard, in equal proportions, one half to each, and I direct that within six months after my decease all of my stocks, not required to carry out the other provisions of this will, shall by my executors be equally divided between my said daughters Alvira H. and Fanny and shall be accordingly made over and transfered to them respectivly.

It is my will, and I do accordingly direct that my said executors whether acting as executors or Trustees under this will, shall not be required to give any bond, either jointly or severally, for the performance of their said duty and trust.

In witness, whereof I the said Augustus G. Hazard have to this my will, and testament contained upon three sheets of paper, respectively numbered 1, 2, and 3. subscribed my name and affixed my seal, upon the third and last sheet thereof, this eleventh day of June in the year eighteen hundred and sixty seven (1867), signed, sealed, published and declared by Augustus G. Hazard, the testator, to be his last will and testament, before the subscribers also in his presance and in the presence of each other and at his request have hereunto set our names as witnesses, this eleventh day of June eighteen hundred and sixty seven (1867).

Augustus G. Hazard

Edward Prickett,

A. D. Bridge

George B. Gordon

State of Connecticut, Hartford County Enfield July 3, 1868

Personally appeared before me Edward Prickett of said Enfield and made solemn oath that he attended the written will of Augustus G. Hazard and subscribed his name in the presence of the Testator and in the presence of the other two subscribing witnesses to said will and that they also subscribed and attended said will in the presance of the deponant and of each other and in the presance of said Testator and that said Testator at the time of the execution of the will aforesaid, was of sound mind & memory, signed, & published said will in the presance of said deponant and the other subscribing witnesses thereto.

J. Warren Johnson
Justice of the Peace

Codicil

I, Augustus G. Hazard of Enfield in the State of Connecticut, being of sound and disposing mind and memory, do make and publish this my Codicil hereby confirming my last will made on the eleventh day of June eighteen hundred & sixty seven, so far as this Codicil is consistant therewith, and changing the disposition of my property in the following manner only, (to wit) Revoking the provisions made on the first and second sheet of said will relitive to my daughter Georgianne and her son Edward B. Sexton, in line thereof I hereby give and bequeath unconditionally to my said daughter Georgianna three hundred and twenty four shares of the capital stock of the Hazard Powder Company, one hundred shares of the Hartford and New Haven Railroad Company, one hundred shares of the Boston and Albany Railroad Company, one hundred shares of the New York and New Haven Railroad Company, and seven thousand and five hundred dollars in money, the foregoing bequest to take the place of the provisions already refered to made in said will for said Georgianne and Edward B. and to be without any conditions of trust or otherwise –

To the provisions made in the second sheet of said will relative to the children of my daughter Emily. (to wit) Anna Mumford, Arthur Hazard, Ellie and Emily, I hereby make the folowing additional bequest (to wit) I give and bequeath to my executors in said will the sum of one hundred thousand dollars in addition to the stocks by said will given and bequeathed from the benifit of said grandchildren, the same to be held and managed by said executors upon the same trust for the benifit of said four children of said Emily, and upon same conditions, as I have by said will provided relative to said stocks bequeathed for the said children of my daughter Emily.

It is my Codicil, and I do direct, that my executors shall hold and set apart and use for the completion of the Hazardville Institute a sum, which, together with that already expended by me for that object and that which I may in my lifetime hereafter expend for the same, shall amount to seven thousand five hundred dollars, and further the sum of one thousand dollars to be held and invested in trust for the preservation and regulation of the property and affairs of said Institute.

I further provide that, in any case in which by my said will or under this Codicil any stocks are specifically bequeathed to my executors to be held in trust, said executors are hereby fully authorized and empowered to make a charge of such investments or any part thereof at any time according to their judgment and discretion.

I further provide that, if, at the time of my decease, it shall appear that by my will I have made a bequest of any particular stock or stocks in excess of the actual amount of such stock or stocks held by me at the time of my decease, then, such deficiency shall be made good by money, to the amount of the make & value of such deficiency of stocks at that time –

In witness whereof, I, the said Augustus G. Hazard, have to this my Codicil subscribed my name and affixed my seal this twenty fourth day of April eighteen hundred and sixty eight.

Augustus G. Hazard

Signed, sealed, published and declaimed by Augustus G. Hazard, to be his Codicil to his will made on the eleventh day of June eighteen hundred sixty seven before the subscribers, also, in his presance, and in the presance of each other, and at his request, have hereunto set our names as witnesses this twenty fourth day of April eighteen hundred sixty eight.

John O. Stone
A. E.. Douglass
Peter C. Cornell

State of Connecticut County of Hartford Enfield July 3rd, 1868. Personally appeared before me Peter C. Cornell and made solemn oath that he attended the within Codicil of Augustus G. Hazard and subscribed the same in the presance of the testator; of the other two subscribing witnesses to said Codicil and that they also subscribed and attended said Codicil in the presance of the deponant, of each other, and in the presance of said Testator, and that said Testator at the time of the execution of the said Codicil was of sound mind and memory and signed and published said Codicil in the presance of said deponant and the other subscribing witness hereto.

J. Warren Johnson
Justice of the Peace

Block Pond

Buggy Accident in Powder Hollow – circa 1900

South Maple Street Bridge

The New Works

The Prickett Correspondence

Introduction: Mary Prickett, a daughter of Thomas and Elizabeth (Smith) Prickett, was born September 10, 1802 in Faversham, Kent, England. She married John Bridge who was born July 3, 1806 in Milton Regis. In 1821 at age fifteen, John had been apprenticed to dredgeman John Gane and his wife Sarah to learn to scrape the Ocean floor for oysters. After seventeen months he was apprenticed to Robert Bridge, his father, for the remainder of his seven years.

He had to promise to "keep his master and mistress' secrets, do their commands gladly, do no damage to his master and mistress, neither waste nor lend their goods; matrimony he shall not contract during the said term, at cards, dice tables or any other unlawful games he shall not play, Taverns, Inns or Alehouses he shall not haunt nor absent himself from his said master and mistress' service day or night unlawfully but in all things be faithful".

In turn, his father agreed to teach him his trade and also give sufficient meat, drink, apparel, washing, and lodging and all other necessaries during the said term. At the end of the term he will give and deliver double apparel of all sorts (that is to say one good new suit for the holidays, a suit for the working days and a pair of boots of a season's wear).

John's father, grandfather, and great grandfather were all involved in the fishing trade in some way, either as fisherman, ship's chandler, or selling fish and oysters. John earned his living as a fisherman but about 1840 he became so crippled with arthritis that he could no longer work and his boat went to pay bills.

Mary had several brothers who had worked in the powder mills at Faversham and had gone to America to work in the powder mills there, in a place called Scitico. They thought the climate might be better for him there and wrote them a letter with this suggestion.

Note: These translations are from original manuscripts collected by Allyn Bridge, MD, are used with Dr. Bridge's persmission and are copied verbatim. Where text was indecipherable, it has been marked with "**** "

"Enfield, June 27, 1841
Dear Brother and Sister,
The last time we wrote we could not see any opening for you to come out but we think there is now if you have any desire to good opportunity. We have been speaking to one of the company of the powder mills he says if you will come and work in the canister house and packing house which is most all setting work canistering powder and pegging up kegs he will set you to work as quick as you get here and give you from 22 to 25 dollars per month which is from 3.6 to 4 today shillings of your money provisions are about the same as the last time we sent you word we think if you ever come you had better come now for the work is easy and little danger the work is such as woman used to do at Halls Mills you had better apply as quick as you receive this if you can get to sail anytime before the first of October you had better come now but if you cannot come now but if you cannot you better wait until next March or April.
"We think you better come in a packet ship because you will have better accommodations come to New York and there inquire for the steamboat that comes up the East River to Hartford and there you will have to leave your goods post them in a storehouse and inquire that runs to Springfield but you must get out at Thompsonville which is eight miles before it reaches Springfield and then you will be three miles from us there go into a shop and inquire for Henry Prickett we have told you but just the heads of the matter but we shall look out for you some

night at the yellow house and then we will tell you the tails you can get your boys to work without much trouble give all our love to all inquiring friends.
With our love and best wishes
W K G & J Prickett"

A letter from Henry Prickett to John and Mary Prickett Bridge in Faversham, England follows:
"Enfield, December 14, 1841

Dear Friends,

I take this opportunity to write to you. We received your letter the latter part of November. We are sorry that you have been so much disappointed by the authorities of your parish. Gentlemen I cannot call them, but we were glad to hear that you were still living notwithstanding all the hard times in the land of the living. God Almighty in his Providence has given you bread. I thank his holy name for what he has done for us both. For our souls and bodies he has given us plenty to eat and drink. We can certainly say he has brought us out of our bondage into a land of milk and honey. ***I shall not say anything about our circumstances for we hope we shall soon see you. Then we can talk. To face ***matters, your mother has gone into that invisible world we all must go, sooner or later. Who knows but our mothers may accompany a convoy of angels if we continue in the faith to the end to conduct our spirits into Abraham's boosum.

We feel interested in the subject of which I am now writing. We spoke to the Company about you; said we thought you could not raise money sufficient to pay your expenses here. We got one of them to write Mr. Hazard in New York to know if there was any way we could send you a little money to help you over. He sent us word it could be done very easy, said he had got several out, but your passage must be paid this side and you must sail from Liverpool. We knew that would not do. We have this evening given Mr. Denslow your directions for him to send to Hazard and he will send you a Bill of Exchange for you to draw 50 dollars at London. We should have wrote before but have been waiting to know about the matter. We thought probably you would want a little help. Whether you do or not, you just draw the money. You may receive a letter from Hazard with the Bill before you get this letter.

Your brother George and myself are responsible for the money. We thought if you did not raise money enough by subscription it would make it so as you could get away. If you do not want it there will be no harm done. We should like you to bring us a dozen gernseys, two Woodmans hand bills and a few pairs of stockings and a three prong digging fork if you can. That is if you have more money than you want have for your expenses. We have a letter today from Silver. He says he shall sail from England latter part of November. I want you to write to Mrs. Reynolds. Tell her we are quite well, that you are coming out, if she can come and see you and bring a letter for me. George wants you to write to George Bushell at Lindal House. Tell him he has received his letter. Ask him to come and see you and bring a few garden and flower seeds, particularly dailyer roots or seeds. Bring me a deck of barley and two or three black currants bushes. ***. We have agreed with Company that you are to work for them one year, and then you can agree for yourself. But we are not afraid but that you will get a living at something or other. He who numbers the hair of your head will provide something for you. I have been young but now old says the inspired writer, but never said of the righteous man "Forsaken", nor his seed begging their bread.

Excuse the spelling if you please.

Henry Prickett

Postscript: Have forgot one thing, Please to ask our dear sisters if they will not send us a small notion of our dear mothers. We want nothing away from them, but just something to remember her."

Included with Henry's letter was the following letter from G. E. Prickett to John and Mary Bridge; Milton, Near Sittingborne, Kent, Old England.

"December 14, 1841"
Dear Brother and Sister,
I take pleasure in taking up my pen to write a few lines to you. I want to ask favour of Mary before you leave England. That is to see if she will go to see Mr. and Mrs. Kempes folks and tell them all the particulars of your coming to America *** and please to tell them and in a special manner to write a long letter and we would be very grateful to you if you would bring us all the tidings about our sisters and families likewise to go and see Mrs. Knowles folks and ask them to write a long letter to us as they will have so favorable an opportunity. I think I am putting you to considerable trouble but I think, my dear friends, if the Lord is pleased to permit you to leave England it will be the last time you will ever see them in this world. And I say to you my dear friends, if you are permitted to make the attempt, may the Lord prosper in your journey to us. *** Please tell our sister Susan that we know she has been expecting a letter from her sister Betty (Kitty?) but you may tell her tis not that she has forgotten them. *** William has taken the Boarding House where she has placed in a very hard place. Tell them, if you please, we ever remember them when we go to a Throne of Grace. Tell them I have had a fit of sickness *** since our last baby was born. Do not enjoy very good health. Tell them please, all the consolation we have from our dear friends that are left in this world is to hear from them. *** We have not heard from our sister Eliza lately. *** Tell them that our sister Betty (Kitty?) will by the blessing of God to write to them after we have heard from them. She cannot write very well and she always wants me to write for her.
I must now leave one and all into the hands of the Lord.
G. E. Prickett

When you get to New York you better change your clothes and make yourselves look as respectable as you can for those steam boats are furnished and carpeted so nice and most of the passengers are gentlemen and ladies. You had better come in a packet as they are safer and quicker. (Actually, they came by team from Hartford to Scitico.) If you possible can bring your bed and bedding and most of your earthern ware but not much else. *** Amen.

Please to inquire if you see George Bushell who John Hills took for a wife and where he works and lives and give our love to him and tell him that we do not want to come back to England.

"New York January 20th 1842
Mr. John Bridge, Sir
Above you have bill of Exchange at sight for ten pounds on Messrs Brown Shipley and Co. Liverpool which we remit by request of Mr. Prickett (in the employ of Messrs Loomiss Hazard and Co. of Suffield Conn.) and the object of which Mr. Prickett has no doubt advised you of.
Yours tr
AG Hazard & Co.
Messrs Denison Hayward, of London will probably cash the above."

In 1845 Henry and Elizabeth Prickett wrote the following letter to John and Mary Prickett Bridge in Scitico.

"Schaghticoke, N.Y., April 21st 1845

Dear Brother and Sister it is now almost a year since we have seen each other's faces when or where we shall next meet I know not god only knows perhaps in Eternity but we can hold a kind of intercourse with each other by Post. I received a note by you by A. Delong the seal of which was broken open before it reached me. I have been waiting thinking perhaps I might hear from you by one of the other men but as I have not I take this opportunity of writing to you. I hope this finds you enjoying both temporal and spiritual health, we are all pretty well. Excepting myself I am a little lame in my right leg last Wednesday night after dark I went to stop the pulverizer the gate shaft being right over the wheel which is a tub wheel and the old timbers being very rotten it broke way and let me down. By the mercy of god my feet fell off the rim of the wheel and my body between the wheel and rock so that I got off without being much hurt. The wheel has a band around it being put on in two halves it consequently leaves ears where it is bolted together projecting over 6 inches from the rim of the wheel these ears struck my leg two or three times while I lay between the wheel and the rock before I could get away so that I have been quite lame but am getting better now.

Had I fell over the rim instead of on it the arms would have taken me to the spout and there have broken me up like a rotten stick from the fire and I to all probability should before this have been numbered with the dead or so injured that I should grieve my life away in misery and "O Dear, O Dear". I thought I was gone as much as cold be the jaws of Death compassed me about but in the Lord will I put my trust. We have had a letter from England, old grandfather Sherwood and so is old Joseph Kingsworth, Mr. Hills the butcher is out of his mind so that he can't be left night or day. We have commenced work here again after a long stop.

Henry is going to burn coal this summer. We have four cylinders like yours he is to have 50 cents per day to burn the four cylinders, it is to be his fournoon's work, in the afternoon he is to work somewheres else. Some difference from John Rooks wages (Thomas Rooks came from England with James and Jesse Prickett in the spring of 1835 and later went to Rochester, NY). Here we burn large wood as large as a man's body from that to his thigh in five hours and as large as your arm in three hours. We make 96 kegs a day that is 26 hundred pounds a day with 7 men and a boy. We had a government order last fall of 541 bbls, we have another order now of 1000 bbls and they are pretty sure of another before the summer is gone we have heard that your government powder has been returned. My. Swift, one of our proprietors was at Boston to see our powder proved. When they had done our powder they tried some of yours. Ourn went 300 from that to 320 yards. Yourn went 150 and some a little more. You run yours 8 hours, we run ourn 6 hours. This is Swift's account of what he saw when in Boston. I have since heard that it has been returned. I want you to write me as soon as you receive this and let me know all about it. Corban brought up some of your Kentucky. I tried it and it goes no better than our Commander that I make with a reduction of petre and 17 pounds of Brimstone to the Hundred, If this is the way you are going to do it you will soon close up shop.

Send me word what your calculations are about building a house and all the news you can – who works the wheels who is Boss and so on and how many blows up you have had since I have been here. I am kind of Boss but not like Veits. I have something to do and see that the work is done beside. We were sorry to hear of another poor man blown up down there. I went to hear his funeral sermon two weeks ago. We have heard that another mill has blown up. Since that I want you to send work if there has. Give our Best

Respects to all relatives and friends.
Henry & Elizabeth Prickett

Next a letter from James and wife Elizabeth, the widow of George Prickett, to John and Mary Prickett Bridge.

"Schaghticoke, April 1847

Dear Brother & Sister,

According to my promise I sit down to write a few lines to you hoping it may find you all in good health and may it thank the Lord it leaves all well. We arrived here safe the same day we left you. After a long day's ride we got to Albany about two o'clock. Had to wait a little while then took the cars for Troy. When we got there young Henry was there with a man and team for to take us to their mills. When we got there it was almost dark and we was very tired, some cold and some hungry and I must say also that we were warmed up and fed and had a bed to rest upon. I commenced to work the first Monday and have worked every day since. It seems to agree with my health very well. So far the work is considerable hard, you know salt peter is very heavy and we have a good deal of heavy lifting, but we get done most times about four o'clock which gives us time for a good rest by bedtime. On the whole I think it is a pretty good job. The place where the mills stand is a pleasanter place than the hollow. It is one mile from Schaghticoke village. We live just upon the border of the town, we have three-quarters of a mile to go to our meeting. The society is small and we do not have preaching but once on the Sabbath. We have prayer meetings in the morning and evening and preaching in the afternoon.

I told you I thought we could live cheaper here than in Enfield. We can a little but not so much as I thought, for provisions are a great deal dearer than last fall. I will give you the price of some things now and what they were last fall.

Mutton was sold last fall for 21/2, beef 4 and pork from 345 to 450, flour 412 ½, corn 45, rye 50, cheese 6, butter 12 ½ and so the price of them now is beef 6,7,8, veal 4,5, pork 7, by the lug butter 20, cheeses 10, flour 8, rye 87, corn 92, wood is from 3 to 4 dollars per cord four feet long. This Spring is very backward here it has been quite ever since we have been here cold; have had as much snow on the ground as at any one time this winter in Enfield. Henry's Betsy talks of coming down to see you this spring but we are not certain as she will if she does, I think she will start in a few days. We have weaned Lucenda (Lucenda, likely one of Elizabeth's children of George Prickett) of her laudanum (laudanum is a tincture of opium). She has not had any for about three weeks and she is getting quite fleshy and is a great deal better than she has been. Gave Henry (Prickett, brother) your letter and understand he has wrote to you. I see by my paper that Mr. Palmer is sent back to Thompsonville another year. I wish you would see him for me and tell him I have not received my back papers that come to Thompsonville after I left as he promised to send them on here. You may tell him that they come safe every week from Boston to this place since he wrote to the editor and would be glad if you would pay to Mr. Owen 12 or 13 cents for the postage of my papers as I came away and forgot it and will send it down by Betsy if she comes.

Please to give our respects to Mr. & Mrs. Wily. Tell them we are well and that the carpet came safe. We laughed about and ask them if the folks who moved into the house that we left see anything of the pair of hand irons. We think we left them in the oven hole. If they are there please to take them to your house until I should have a chance to get them here. Betsy has wrote to Catherine and to also Thomas Rook. You may have heard of that and began to think I had forgot my promise. I did not think it would be well to write until we had been here some little time. Give our respects to all inquiring friends, to Mr. & Mrs. Locke and her sister, to Mr. & Mrs. Kitson, to Mrs. & Mrs. Brackenton, to all who inquire after us when you are round with your emtions. (sic). Please to answer this as soon as you can and give us all the news you can. Tell George (the eldest son of John and Mary P. Bridge) please to send me news about the shops (George must be working in the carpet mills at age 16 before going to New Haven for apprenticeship).

Also, we know that James worked in carpet mills before power looms. (Letter from James and Elizabeth Prickett to John and Mary Prickett Bridge, April 1847) whither they have as many shop meetings as ever and whither the good old Constitution hold good and whither there is any signs of the power looms to be put in operation and whither they have any ore per yard than they did when I left there and tell George to give our respects to John Fallows senior and to William Taylor and to all the shop mates that inquire after us. When you write please to send us word whither you have seen anything of Jane since we left. Betsy has wrote to Mrs. Abbe but has not received any answer. Send us word whither John W. (John & Mary P. Bridge's second son. He died age 20) still works in the New factory and all the news you scrape together; fill a sheet full as I have done. Give our love to Kitty and tell her that we received her letters and was glad to hear the good news of Edward and Thomas. I must now conclude as I have no more to say this time.

With our love and best wishes to you all,
James and Elizabeth Prickett"

A letter (provided by Norman Edward Henry) from Henry Prickett to John and Mary Prickett Bridge reveals the deep Christian faith of the Prickett family.

"Schaghticoke, Nov. 20th, 1847

Dear brother and sister,
**** The account that I now have to give you quite weighs me down. I mean an account of the sickness and death of my wife. I suppose you have received a letter from James, I asked him to write you a few lines, which he did, but I thought I would write you anyhow. I said in my last that my wife was pretty comfortable a week after her confinement (October 25, 1847), but she did not gain strength as she had done before. On the fourth at two or three in the morning she was taken with a very severe pain. The doctor bled her, easing the pain. He said the liver was affected and that the inflammation was got in and must be broken up. He attended her twice a day, bled and blistered her continually. They concluded however that taking the course of medication they prescribed she would get better **** but she died on the 10th. ****after calling on the Lord to receive her spirit **** I pity myself and the children, not her. Her body is intured in a quiet little graveyard long side the river where I intend to raise **** to her memory a nice large white marble stone on which may be seen 'Elizabeth, Wife of Henry Prickett, Died Nov. 11th 1847, Aged 38 Years".
Dear friends, This is a hard trial, a sore affliction but the Lord saw it was best or he would not have taken her away from us; we can't see according to our short sightedness how it was best to take her who has been the partner of my joys and sorrows for nine or ten years and leave seven motherless children to mourn her loss. Joseph made a mourning for his father, 'The children of

Israel wept for Moses 30 days, so the days of weeping and mourning for Moses were ended but I fear even after 30 days we shall mourn for her who has ministered unto us in sickness and in health." We had a very large funeral, a procession of carriages and wagons followed a meeting house full.
Many dropped a silent tear as they passed along the alter after the sermon to take a look at the life like clay. It was an affecting scene to see me and the children hand over the once Loved Object of our affections. Almost all were melted in tenderness for us. We discovered after she was dead it was mortification (gangrene) that hurried her off. ****I suppose all was not right within after the last child, Emma was born. I have put her out to a woman (wet nurse). Son Henry *** is so lame that he has had to stay at home. The other children are well ***. If I could keep her who is forever gone *** The Lord's will be done, not mine. The Lord gave and the Lord taketh away, blessed be the name the Lord. ***
From your affectionate brother, Henry Prickett"

The following letter was written by John Bridge, the first of the Hazardville Bridges, to his son George, who is a blacksmith apprentice in East Haven, CT. John is 42, George is 18. The '49 California gold rush is on.

"Thompsonville, March 19, 1849

Dear Son,

*** We admire and hope we shall appropriate the hint of caution given to 'Think twice before we jump once'; but allow me to indulge a few moments on the subject of removing. *** You know the primary object of all is the bettering of the present condition, and although this is our object, yet I believe we have a different end in view to many. There are many people in the world whose aim is personal aggrandizement, rich honor or fame. *** Many will give up their business, part with their families, all that is dear to them, and go to California, thinking to gather gold as in the street, others will rush to the battlefields, risk their lives to win laurels that will redound to their honor, and others will exercise all the power of their minds to out some new invention or to settle some great question that their fame might be spread throughout the length and breadth of the land, but neither of these hit our case. I believe our only object is to get a living as comfortable as we can and pay everyone their just do. In order to do this it requires that we be diligent in business and if in the providence of God we must look for another in order that we may procure the supplies we need for our support through times.

"You seem to think we should not better our condition if we move down there; I thought we might but I do not now. I see a very poor prospect here, the village is in a very bad state. Many of the weavers have been out from 15 to 18 weeks, and some few are gone, but let them go wherever they will, there will be no chance for our little boy, for perhaps you can recollect what a number of little ones there used to be running around here. *** I believe the Brussels weavers have more children able to draw than will *** get but six dollars instead of eight.

"The way I thought we might do down there is this, get son, John a place where he would receive wages and board at home which would be considerable toward house keeping, and most likely I should be able to earn a little fishing, and that would make the employment of the little boys (Amos, Ephraim, Ebenezer, and Stephen) (son John W., 17, works in carpet manufacturing) if we could not let them out, perhaps I might earn more in my way of fishing than I expect, *** but there is a great many shrimp there the latter part of the summer and they never catch any to send to market. I would certainly try it and if I could not sell any about there I would send some to Thompsonville. *** I have a little shoe mending (he cobbled in England) here in winter but none in summer and that would probably be the case. If so, I might have some employment there in summer as well as winter; I quite believe we are 30 Dollars worse off than we was 6 months ago, and I do not care to spend the carefully earned little without a further effort.

"These are my reasons for wanting to remove. I know it is a considerable undertaking to remove with a family so large as ours, but the old proverb is "What can't be cured must be endured", so we say if we cannot get any place to do better we must do the best we can here or anywhere else. I hope we shall endure all things with a Christian spirit.

"We received a letter from Uncle (Prickett) this week. *** The principle of his letter is the folk gone west. It appears the farm they bought is in *** Richmon, County McHenry, Illinois. His children, Henry and Mary Ann's husband have been teaching school. The winter took possession of the farm this month. ***They like the country first rate, it is considered a healthy and pleasant county, not much winter. They have had not sleighing there for six years. *** "Johnny, if you was there and settled with 50 acres of fine prairie land I think you would enjoy yourselves right well."

*** He then describes his farm *** (which includes 140 good grafted apple tress, 500 more young grafted apple trees. "When our folks got their wheat at the highest, $1.17 per bushel, corn 31 cents, oats 25 cents, but wheat is down to 70 cents now because of navigation being closed. Beef is now (mid-March) two cents per pound, pork two cents, butter 12 ½ cents, Mutton you can have for the picking. They only raise sheep for the wool.

"Groceries are about ten percent higher there than here. *** Many will fetch good interest, about 30% is the average. Game is plenty. Rabbits, prairie hens and quails. He says likely he will come down again before long. I would remark here, you spoke about coming up for the 4th of July, but I think I would hold on. I shall know very soon whether he will come or not, and when he comes we should very much like you to come up then. *** His time is out in Schaghticoke this May. He does not know whether he shall go west this summer. It appears Uncle James Prickett has been very sick indeed with the California fever. Henry says "He was all on fire for it as much as any man. He has concluded not to go, but we have sent a man which cost us $200, $100 each. We are to have half of what he gets when he comes back. He has got his life insured for $500 and given us the policy. James thinks if this man does well he shall be able to go west and get himself a little farm." *** I hope you are doing well with the net. Write again as soon as flat fish come in plentiful. *** I think I shall come down then and try and catch some. I think if I had a net I could pay my expenses by catching a lot of fish and bringing up with me. I expect the flat fish will very soon be in. Learn all you can about fishing and let us know.

J. & M. Bridge

The following letter is from John Jr's brother George, 19, to his parents.

East Haven, Aug 3d, '49

Dear Mother and Father,

I received yours dated July 17th and was very glad to hear from you. I delayed writing till now because Thomas and Edward came up. You say you expect I find it pretty hot *** I like the warm weather very well I have been pretty uncomfortable three or four days here, One day the temperature was 99 to 102.

*** I did not sleep very well that night. I am getting along very well at my trade, especially horse shoeing. Last Friday we did not work. I went to church twice (and Mr. Holts at noon) Uncle Henry (Prickett) has gone west. I should liked to have seem him before he went but perhaps we shall all see him where he is before long. There is no less than five young ladies from (Thompsonville) here now. They say they are stopping a good many looms. I hope John won't have to stop How does your garden get along. Hugh Galbreath sends his respects. Direct: Blacksmith Shop, Bridge Street.

I am your Affectionate son, Geo. Bridge.

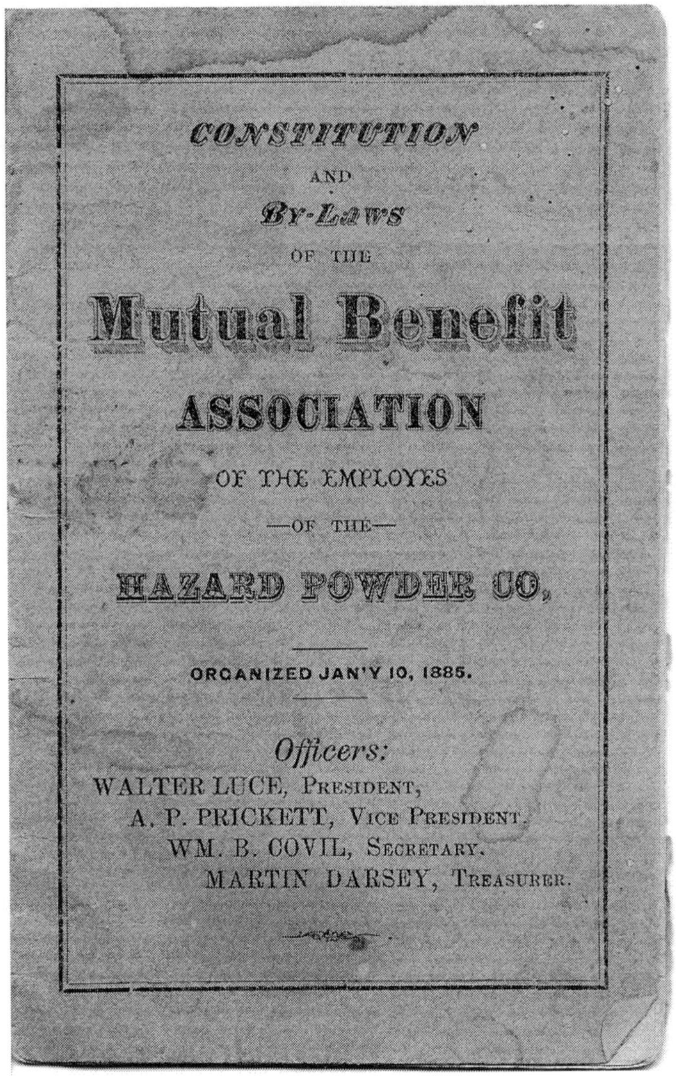

CONSTITUTION

AND

By-Laws

OF THE

Mutual Benefit

ASSOCIATION

OF THE EMPLOYES

—OF THE—

HAZARD POWDER CO,

ORGANIZED JAN'Y 10, 1885.

Officers:

WALTER LUCE, President,

A. P. PRICKETT, Vice President.

WM. B. COVIL, Secretary.

MARTIN DARSEY, Treasurer.

CONSTITUTION

And By-Laws

of the

MUTUAL BENEFIT ASSOCIATION

of the Employes

of the

HAZARD POWDER CO.

Organized Jan'y 10, 1885

OFFICERS

WALTER LUCE, President

A. P. PRICKETT, Vice President

WM. B. COVIL, Secretary

MARTIN DARSEY, Treasurer

CONSTITUTION

ARTICLE I.

NAME.

This Association shall be called the MUTUAL BENEFIT ASSOCIATION of the Employes of the Hazard Powder Co.

ARTICLE II.

OBJECT.

The subject of this Association, shall be to render financial aid to its members, in case of sickness or to their representatives, in case of death, as hereinafter provided.

ARTICLE III.

OFFICERS.

Sec. 1, The Officers of this Association, shall be a President, Vice President, Secretary, Treasurer, and three Stewards.

Sec. 2. The President, Vice President, Secretary, and Treasurer shall be elected by ballot, and hold office for one year. The Stewards shall hold office for three months, and be appointed in rotation, as their names appear on the List; provided that those shall be excused who hold other offices.

ARTICLE IV.

DUTIES.

Sec. 1. The President shall preside at all meetings and preserve order.

Sec. 2. The Vice President shall preside at all meetings, in the absence of the President.

Sec. 3. In the absence of the President, and Vice President, the meeting shall be called to order by the next higher officer present, reckoning the Stewards, as their names shall appear on the Roll, when the meeting shall elect a presiding officer, pro. tem.

Sec. 4. The Secretary shall keep a record of the doings of the Association, in a book provided for that purpose. He shall also keep a record of the officers present at each meeting.

Sec. 5. The Treasurer shall be the custodian of the funds of the Association, giving bonds for the safe keeping of the same. He shall keep a record of all the financial transactions, in a book provided for the purpose, which book shall be open for the inspection of the Association, at all its regular meetings.

He shall make a report quarterly, of all the money received and paid out, and no money shall be paid out by him, except by vote of the Association, and the bill be approved by the President and Secretary.

Sec. 6. It shall be the duty of the Stewards, upon notification that any member is sick, to visit such member at least once a week, by one or more of their number, while he continues sick, and report to the Association at its next meeting, provided; that if any Steward shall himself be sick, the next on the list, shall serve in his stead.

ARTICLE V.

ELECTION.

The election of officers, shall take place at the annual meeting, in January of each year, and any vacancy that may occur, by reason of death, withdrawal, or expulsion, may be filled at the next quarterly meeting.

ARTICLE VI.

MEMBERSHIP.

Sec, 1. Any person in the employ of The Hazard Powder Co., shall be eligible for membership, provided he has no chronic disease or disability.

Sec. 2, All persons desiring to join this Association, shall make application in writing, through some member, which application shall be endorsed by two members in good standing, and shall be acted upon at the next regular meeting following, and if two-thirds of the members present shall so vote, the applicant shall be declared a member, upon his subscribing to the Constitution and By-Laws of the Association, and paying the initiating fee.

Sec. 3. Voting for members shall be by ballot, "Yes", or "No".

Sec. 4. Any member leaving the employ of The Hazard Powder Company, shall still retain his membership, so long as he continues to pay his dues, and if he shall move out of Town, may receive aid, by furnishing satisfactory evidence of sickness.

Sec. 5. Every person, at the time of joining this Association, shall designate to whom the money shall be paid, in case of his death.

ARTICLE VII.

MEETINGS – QUORUM.

Sec. 1, This Association shall hold regular meetings, at least once a month.

Sec. 2, Special meetings may be called by the President and Secretary, three days notice having been given, by notices posted by the Secretary, in the Post Office. Hazard Powder Company's Office, and in each watch-house.

Sec. 3. Eight members in good standing, shall constitute a quorum, for the transaction of business at any meeting of the Association.

ARTICLE VIII.

SICKNEES – AID.

Sec. 1. In case a member is sick, or injured so that he cannot work, he shall be entitled to Eight Dollars ($8.00) per week, up to eight (8) weeks, provided however; that no member shall receive more than Sixty Four Dollars, ($64.00) for any one sickness, or during any one year, and also provided; that no member shall draw any pay, for any sickness of less than one (1) week's duration.

Sec. 2. In case of the death of a member, there shall be paid to the person designated by him, the sum of Thirty Five Dollars, ($35.00).

Sec. 3. Upon a member being taken sick, he shall at once notify the Stewards, and no member shall be entitled to any aid for the time he was sick, previous to such notification.

Sec. 4. Any member who shall let his monthly dues run over for more than one month, shall not be entitled to any aid during the time, and any member neglecting to pay his dues for three months, shall lose his membership.

Sec. 5. Any member making a false statement, in order to defraud the Association, and procure aid when he is not entitled to it, shall be expelled from the Association.

ARTICLE IX.

CONTINUANCE OF ORGANIZATION.

This Association shall preserve its organization, as long as Ten (10) members require its continuance, unless two thirds of the members united in a resolution to that effect, and notice having been given, at a previous meeting.

ARTICLE X.

WITHDRAWAL.

Any member may withdraw from the Association, if his dues are paid, by giving notice in writing, at any regular meeting, which notice shall be a relinquishment of all claims upon the funds of the Association.

ARTICLE XI.

ALTERATIONS – AMENDMENTS.

This Constitution may be altered, or amended, by a two-thirds vote, at any quarterly meeting, notice having been given in writing, at least three months previous, and this notice shall be read at each meeting, until its adoption, or rejection.

AMENDMENTS

I.

Article XI of our Constitution, shall be Article XII.

Voted Jan'y 8th, 1887

II.

SUBSTITUTE FOR ARTICLE XI.

All persons making application for membership into this Association, and receiving an affirmative vote, shall present themselves to sign the Constitution and pay their fee – becoming members in full – within three months of their application.

Voted Jan'y 8th, 1887

III.

ADDITIONS TO ARTICLE VI.

Sec. 6. When a person withdraws from this Association, the Treasurer shall write the date of such withdrawal, opposite his name, on the books, and such person shall not be proposed for membership again, in the Association, within one year.

Voted July 9th, 1887

IV.

AMENDMENT TO SEC. 3. ART. VIII.

Upon a member being taken sick, he shall at once notify the Stewards, and no member shall be entitled to any aid, for the time he was sick previous to such notification; and when requested by friends of the sick person, the Stewards shall provide watchers for the nights, from members of the Association commencing and going through the List of members alphabetically; but no one shall be called upon to perform this service more than once, until the whole List of members have been called upon, and if the whole List is not called upon, during the sickness of any one person, then the next time such services are needed, the Stewards shall commence on the List, where they left off in the previous case, Stewards furnishing this information to their successors.

Voted July 9th, `1887

BY-LAWS.

I.

Every person joining this Association, shall pay an initiation fee of one dollar. ($1.00)

II.

All members shall pay monthly, the sum of fifty cents, (50 cents) into the Treasury of the Association, except as hereinafter provided.

III.

Any member who shall be sick for more than one month, shall have his dues remitted, for the time he was sick.

IV.

No member shall receive any aid from this Association, whose sickness shall have been caused by fighting, drunkenness, or the venereal disease.

V.

No intoxicating liquor, shall be brought into any meeting of this Association.

VI.

The regular meetings of this Association, shall be held on the Second Saturday Evening of each month.

VII.

These By-Laws may be altered, or amended, by a two-thirds vote, at any regular meeting, notice having been given in writing, at least one month previous.

Horseshoe Falls and Cylinder Mill

Bridge over the Scantic in Powder Hollow

Scene from The Hazard Powder Company

Railroad Building

Remnants of Graining Mill – June 18, 1897

From a postcard published by G. E. Allen

Powder Hollow, February 26, 1910

Explosion at Press Mill – September 19, 1896

Canal in Powder Hollow

Picture of dry canal – December 5, 1898

Another working canal – December 5, 1898

Long Hollow Road – August 21, 1899

New works and brimstone coal shed - Powder Hollow – 1877

Coal Shed and Office in background – 1877

Addendum

By Rev. Craig Peel - 2011

From our convenience-filled twenty-first century vantage point, it is all but impossible to overstate the level of faith, courage and sheer tenacity demonstrated by those immigrant forbears who arrived in the tiny village of Hazardville, Connecticut, in the early 1840's! They brought with them high hopes as they embarked upon a deeply uncertain future to accept the invitation of an unknown entrepreneur in the manufacture of explosives.

Being people of faith, they hung their hopes on words of Holy Scripture that up to this point may well have simply been the Vicar's words on the Sabbath. Bearded Abraham, once a distant Biblical hero, likely took on a whole new personality. His call from God as he heard the whisper of God's voice likely morphed into theirs: "Go from your country and your kindred and your father's house to the land that I will show you…." (Genesis 12:1) But, as we move out in faith, will there really be a job? What about school for our children? Is this move permanent? Will I ever see my kinfolk in England again?

From the moment they set foot on their new land, life for them became not only a newfound joy, but also a challenge which pressed them more deeply into the faith they had brought with them! In those not so infrequent moments when life asked of them a deeper courage than they thought they had within them, they must often have counted on the truth of the words of Scripture.

One can only imagine the level of anxiety confronted by wives repeatedly sending husbands and sons off to work in the Hazard Powder Company. Let alone the quietly internalized fears the men themselves took to work with them.

In those breathtakingly devastating moments when a powder mill blew, and the village was turned upside down with grief, what comfort these words must have been: "The Lord is my light and my salvation, whom shall I fear? He shall hide me in His shelter in the day of trouble…" (Psalm 27:1&5).

What a gift the little Methodist Church must have been to these resilient pioneers when it opened its doors in 1835. It must have seemed almost too good to be true to have a place, a building, set aside for worship and fellowship. At last, a place to take all those private, personal thoughts and prayers, fears and hopes into a caring community of faith.

Not so long after the ground-breaking for the Methodist Church came the opening of St. Mary's Episcopal Church for the Anglicans among those early settlers, and St. Bernard's Church for those who were Roman Catholic.

In the pages of this little slice of post-colonial Hazardville, you have been privileged to meet and to come to know many of those forward looking, brave, faith-filled pioneers whose courage and perseverance is truly remarkable.

We are blessed to live in a world which those early settlers could not possibly have imagined. A global world, in which nations and continents are taking on

dramatically new shapes. A world marked by inter-connectedness, the expansion of technology, great promise and opportunity, as well as awesome challenges.

In the face of the hope and challenge of this new day, we have done well to stop and linger, to visit with, to meet and to come to know some of the very special people who forged a new life, and built a strong community near the banks of the Connecticut River as it wound its way from the southern Massachusetts border into Connecticut early in the 1800's.

There has been much to learn from the life they shared and the way they lived it.

VIGNETTES OF OLD HAZARDVILLE

Early Physicians of Enfield

The following account of the Early Physicians in Enfield, by Dr. John L. Bridge, has been reproduced here using the same language, spelling and abbreviations as in the original papers.

Edward and Alice Allen owned Dr. John's house at 23 School Street in the 1970's and when it was subsequently sold to another party, the new owner found this handwritten document in the eaves of the attic and returned it to the Allens. The original papers have been given to the Bridge Family.

Dr. Ebenezer Terry was born on Enfield St. in 1696. His grandfather, Samuel Terry, a citizen of Springfield, Mass., was one of the original proprieters who laid out the Town of Enfield and was allotted a piece of land near the house of the late Robt. F. King. The above mentioned Samuel Terry never settled in Enfield, but his son, Samuel Terry, Jr., was allotted the land originally laid out for Samuel Terry, Sr., "having occupied it since the settlement of the town". Samuel Terry's house was standing in 1829 and must have been built before 1730, the date of his death. Dr. Terry's mother was a next door neighbor of his father's. She was Hannah Morgan, a daughter of Isaac Morgan, one of the original proprieters of the settlement in Enfield. Isaac Morgan was a prominent person in the settlement and evidently a man of vigor and enterprise [and,] In 1675, a soldier in King Phillip's War.

Interested in several business projects in the town; hauled to court in Northampton "for unnecessary travel on the Sabbath Day" and finally drowned in the Conn. River, Nov. 23rd, 1706 at the age of 56. The oldest stone in Enfield Cemetery dated 1696 is "for son of Isaac Morgan".

Marriage of Samuel Terry to Hannah Morgan in 1683 was the first marriage after the settlement of Enfield. Samuel Terry was also a prominent figure in the town, buying and selling land, engaging in business projects, and holding town offices.

Dr. Terry probably remained in Enfield until 1717 when he was 21 for we find his name as witness to a deed drawn up in July, 1717. Later he went to South Kingston, R. I. where he undoubtedly studied medicine. Here he married Mary Helms, great granddaughter of Sergeant Christopher Helms of Warwick, R. I. who died in 1650.

At a town meeting held Mar. 21st, 1722, it was "voted to give Ebenezer Terry 30 pounds to come & keep schools here in this town for one whole year next ensuing. At a meeting of the proprieters in Enfield held Mar. 22nd, 1722 it was voted to give Ebenezer Terry 50 acres of land somewhere between the town and Scantick - - in case he comes and settles here within one whole year next ensuing

& continues here seven years. At a proprieters meeting held on May 13, 1723 it was voted "that ye commoners do now take ye forfeiture Doctor Terry's alias Ebenezer Terry's grant of 50 acres – he not coming according to ye time prefixed in ye grant".

On Oct. 2nd, 1722, Nathaniel Horton & wife deeded to Ebenezer Terry of Kingston, R. I. 5 ½ acres home lot & house on Enfield St. The lot on the N. E. corner at Baker's Corner now occupied by the Hopkins family was allotted to Nathaniel Horton in 1690 and was the property sold to Dr. Terry. Evidently, Dr. Terry settled in Enfield and began the practice of medicine in 1722 or 1723 living in the above mentioned house. Some of Dr. Terry's children were born here including his daughter Penelope.

About 15 years ago the late Dr. Edw. F. Parsons told me that he met a gentleman by the name of H. H. Terry of Springfield, who had one of Dr. Terry's old account books. The first entry in this book was made (just 200 years ago) in 1725. H. H. Terry added that he bled everyone in town. His charge was 1 shilling for bleeding. Dr. Terry did not remain very many years on Enfield St. for we find him acting as school committeeman in Wallop in 1736 and in the town records of 1743 it is stated that a road led past Dr. Terry's in the eastern most part of the town. Dr. Terry mortgaged his place on Enfield St. to John Pynchon of Springfield in 1728 and again in 1730.

Sometime later it was turned over to John Pynchon. Probably, Dr. Terry moved to Wallop between 1730 & 1736. The homestead of the late Albert Terry of Wallop was probably the home of Dr. Terry. Three men by the name of Christopher Helms Terry, father, son and grandson were probably the connecting links between Dr. Terry and Albert Terry. Dr. Terry's son Christopher Helms was born in 1736, his son in 1766 and the 3d about 1808 as he was made an elector in 1829.

Dr. Terry was evidently an educated man otherwise he would not have been hired to teach school. One finds his name cropping out in town records during his life time, witnessing deeds and with caring for the town poor, hiring teachers and paying them, seating people in church and negotiating with ministers. He held many offices in town and at one time represented the town in the General Court of Mass.

This town was well settled when Dr. Terry began his practice here, although it was laid out only 40 years before this time. People were cutting their farms out of the woods in all parts of the town. There were a good many roads but those roads were not what we see today, even in country plans, many were trails through the wood among stumps. Bears, wolves and panthers were found in the east part of town. Most travel was done on foot or on horseback as light wagons or pleasure wagons, as they were first called, were introduced between 1810 – 20.

Drs. carried their medicines in saddlebags and had to make their calls wet or dry, hot or cold, on horseback. It was no easy matter to call a doctor in those days. If a patient lived 5 miles away, the one who called the doctor had to travel 10 miles. Probably they put off the call as long as possible. Undoubtedly, Dr. Terry practiced among poor people all of his life, some of them very poor and living under discouraging environments. Farmers had practically nothing to sell. They needed nearly all the grain they could raise and cattle were few & of poor quality. There were no manufacturers. There was little money in circulation, and the Dr. had to take his pay (if he got any) in produce.

Dr's. were not equipped as they are today by any means. They did not even have any anesthetics. Operations had to be done with patients fully conscious. A few of the drugs now in common use were then known but only a few. Fortunately opium was known and used. For almost all kinds of diseases the patient was bled. This custom continued until 1840 or later. Now people are bled by doctors but in a different way and no one gets off for a shilling.

In spite of all these handicaps these men of the old times brought lots of comfort to the sick and feeble. They could open abscesses, set bones and relieve pain. Dr. Terry was in a class by himself. He was the pioneer doctor of the town and served the people under the most difficult conditions. Probably for the most part of the first 40 years of his practice he was the only doctor here. There may have been a few years before 1760 when another native doctor practiced in Enfield.

In 1760 Dr. Daniel Dwight, who served in the French & Indian War, was a son of Enfield parents and a married man who went with one of the expeditions to Canada and died at Oswego, N.Y.

During the year 1763 a new doctor came to town. At that time Dr. Simeon Field, son of Thomas Field of Hatfield, Mass., settle in Enfield. Dr. Field was born in Longmeadow in 1731. About the time he came to town he married Margaret Reynolds, a young woman of 20, daughter of Rev. Peter Reynolds, pastor of the church here. I believe he occupied, during part of his life, the house lately owned and occupied by Wm. Abbe, corner of Bridge Lane. Dr. Parsons told me he lived here and said he had his window sign. Dr. Field and his wife had 3 sons and 2 daughters. Two of the sons were physicians. Simeon Field, Jr., and Edward Field, for whom Dr. Parsons was named.

From the beginning of Dr. Field's, Sr., residence here he was prominent in the church. He served as clerk & treasurer for many years. He is also mentioned in the town records many times, holding offices and taking active part in town affairs. While Enfield St. was a better place for a doctor's office than Wallop and conditions were better in 1763 then in 1723 when Dr. Terry began practice, Dr. Field had to serve people living under unfavorable circumstances. Roads were very poor and calls had to be made on horseback. Until after the Revolution the inhabitants were poor and in a discouraged condition.

We had good evidence that Dr. Field was an up to date doctor as well as a public spirited citizen. For several years he was engaged in a fight to have the town undertake the establishment of a station for the inoculation of smallpox on the island in the Conn. River called at that time, 1788, Copper Island. Inoculation consisted in vaccinating a person in the arm with the discharge from the pustules of a smallpox patient, actually giving the person smallpox. The disease given in this way usually ran a short, light course, but protected from smallpox as well as smallpox contracted in the usual manner.

Dec. 1st, 1788, the town voted to inoculate and Jan. 12th, 1789, this vote was rescinded after many attempts to induce the town to inoculate. In 1809, one year after Dr. Field died, they voted to try it for 3 months but even this vote was rejected 2 months later and a substitute to permit inoculation passed. About this time vaccination became more common and inoculation became a thing of the past.

While Dr. Field was an old time physician, he died in 1801, Enfield St. probably did not look much different than it did within the memory of people still living. The old church (now the community house) was built 12 years after he settled here. Several of the houses now standing in town were here in his lifetime.

Two other doctors should be mentioned as physicians of the older time. Having died over 100 years ago. Dr. Simeon Field, Jr. was born in Enfield in 1765. He graduated from Yale College in the regular course and was afterward granted the honorary degree of M. D. by Yale. He practiced medicine for a time in Somers but later came to Enfield, possibly near the time of his father's death. His mother's brother, Dr. Samuel Reynolds, graduated from Yale in 1750 and practiced in Somers until 1774. Dr. Field, Jr. was prominent in town affairs and was moderator at town meetings for several years. He died in 1822 at the age of 57, the record says of "fits".

Living not far from Dr. Simeon Field, Jr. and practicing at the same time, was Dr. Solomon Bond. Dr. Bond built and occupied the house just south of the school in Enfield St. later owned and occupied by Dr. Horatio A. Hamilton, now owned by the Woodward family. Dr. Field and Bond were men of considerable wealth and prominent men in the town. They must have been friendly during a portion of their lives for Dr. Field witnessed Dr. Bond's will dated about 3 weeks before he died in 1814. Dr. Field's will was dated about the same time.

These men practiced medicine under more favorable circumstances then Dr. Terry. Medicine had become more of a science and a few medical schools had been started.

I mention Dr. David Allen, who at one time owned a good part of "Powder Hollow". I can find very little written about him. He probably was a soldier in the Revolution. I know he was receiving a pension from the U. S. govt. at the time of his death in 1834. He was a man of considerable wealth.

We have in the church records from 1808 to 1849 a list of names of persons dying in the town. Ages are given and in many cases the disease of which the person is suppose to have died. What strikes one in looking over these records is the comparatively large number of young people mentioned. Of the first 50 whose names are recorded, 20 were under the age of 20 and over half were under 40. There were a number of severe epidemics of children's diseases.

During a six months period 20 children died of scarlet fever. The population of the town was at that time not more than 1/7 of what it is today. Many died of a disease called "rattles" which was undoubtedly laryngeal diphtheria, at one time called membranous croup. A very large number died of consumption and those were mostly young people. It is then easily understood why the names of so many young people were found in these records.

In a good many cases you have to read between the lines to learn the real cause of death. For instance, during one autumn 7 Shakers died of a fever, when 7 people living together die of a fever in the fall that means typhoid, which was carried from one person to another by food or drink.

Some curious diagnoses were made. A very large number of people died of languishment. This probably means that many people had disease in which there was a gradual failure of strength. Several diseases have this noticeably. Tuberculosis, cancer, diabetes, anemia and nephritis. When the diagnosis of consumption appeared, languishment became less common. Another number died of what was called "fits". The ages of those people ranged from infancy to the oldest inhabitant. Why they called the cause of death of so many "fits" is hard to understand. Probably a convulsion of any kind was called a "fit".

We can have these at any age and from a good many cause. In infancy and childhood they are common. At any time of life they accompany serious kidney disease. I find too, that any attack of sudden unconsciousness was called a fit. What we call "shocks" would come under this head. Also sudden death from heart disease. We ask who made these diagnoses? Certainly not the person who kept the church records. The doctors who practiced 50 years before these names were recorded taught the people the medicine the doctor knew at the time. The two Simeon Field and Dr. Bond were responsible for these diagnoses. They recorded what they observed and we can do no more.

Written by Dr. John L. Bridge (circa 1925)

The Mansion known as Fairlawn

Colonel Hazard built Fairlawn for his daughter who was to marry Mr. Garische, a foreman with the Hazard Powder Company. Garische was killed in an explosion and never occupied the mansion. Subsequently, Thomas Sexton married a second daughter of Col. Hazard and they lived in the home which still stands north of the playground on Main Street. Edward Prickett ultimately became Superintendent of the mills and lived at Fairlawn with his family until his retirement.

Years later Fairlawn was in a state of disrepair.

Now under new ownership Fairlawn has been renovated without its tower

Fairlawn in winter

Men Principals of Hazardville School

Hazardville School

Men principals in Hazardville School beginning after 1862 where present school was built. There were women principals 1861 and 1862. The first principal I remember was Mr. Bronson, a mild mannered man. He was principal probably about 1867. Next came Mr. Philip Pearl, 1868-70. He married the daughter of Miles Spencer of Scitico and had at least 2 children – Jean and Spencer. He was principal twice and later was principal of a grammar school in Hartford. After Mr. Pearl's first stay came Mr. Dutton from Vermont who afterwards became a doctor and practiced in Stafford Springs, I believe. Then came John McKinney, who came from Suffield Academy, after being a student there. Then Mr. Philip Pearl's second stay. He left in the middle of a term because the pupils were so bad.

Then came Mr. Coates, who became a lawyer later and practiced in New Britain. John S. Cooley, 1873 to 1875, then became principal. He was a stout man. Also was a fine penman and taught penmanship to the teachers of Enfield. He left teaching and traveled for a school book company. He called on me in 1917 as an agent for school book publishers.

After Mr. Cooley came Mr. Adams, a graduate of Williams College and the only college graduate to teach here up to his time. He was a fine principal and taught many things not in the school text books. Soon after Mr. Adams came Mr. Elmer Randall who was principal here for many years.

Written by John Bridge

History of E. C. Allen's

The two-story, wood frame building was built in 1850 by John Bridge, the father of Amos D. Bridge who was the founder of Amos D. Bridge's Sons, Inc. In 1875 Edward C. Allen, now a representative to the general assembly, accepted a position in the store, and served in that capacity until 1881, when, with A. W. Gowdy, he embarked in general merchandising under the firm name of Allen & Gowdy. This partnership existed for only one year, at the end of which time Mr. Allen purchased Mr. Gowdy's interest. Upon the death of E. C. Allen, I, his sons, George and Herbert took over the operation. George worked until his death in 1955 and Herbert worked until his death 1976. George's son Arthur and Herbert's son Edward took over the business until closing in 1993.

The building was subsequently sold to the Methodist Church whose property is adjacent to the store property. It is currently rented as a retail operation.

E. C. Allen's original building. The family lived in the apartment upstairs until moving into their new home across the street, about 1888. The store and the apartment were not occupied at the time of the fire. E. C. Allen is standing on the left with his son George.

When fire destroyed the building in January, 1919, E. C. Allen moved his business across the street to the building now occupied by Atty. Charles Parakilas. E. C. Allen's business continued there until the 'new' store was built in 1930 on the site of the original building.

Scenes of E. C. Allen's store after the fire

Edward C. Allen, I with his grandson Edward C. Allen, II (1928)

Allen's Store Fire

This is a handwritten account of the discovery of the fire which destroyed the store owned by E. C. Allen, I in the village of Hazardville. Written Jan. 6, 1919, author unknown.

E. C. Allen's store, a landmark in Hazardville, burned to the ground last evening. It probably carried one of the most complete stocks of general merchandise entailing a loss estimated at about $30,000. It was erected in 1850 by John Bridge, the father of Amos D. Bridge who was the founder of Amos D. Bridge's Sons, Inc.

In 1875 Edward C. Allen, now representative to the general assembly, accepted a position in the store, and served in that capacity until 1881, when, with A. W. Gowdy, he embarked in gen'l merchandising under the firm name of Allen & Gowdy. This partnership existed for one year, at the end of which time Mr. Allen purchased Mr. Gowdy's interest.

The fire was discovered by Mrs. George W. Campbell, who was walking past the store about 20 minutes of nine and heard a dull explosion, followed by smoke. She ran to the engine house, where the volunteer fire company were holding their annual meeting. When the company ran from the engine house the store could be seen in flames, and although they put four streams of water on the building, using the new steamer it burned to the ground. At midnight the firemen were still playing water on the ruins. A delegation from the Thompsonville Fire Department assisted the local department. It was thought at first that the barns in the rear, and the residence of Mrs. Henry Stimson, which stands less than fifty feet away, were in grave danger, but the good work of the fireman prevailed.

NOTE: per Alice Godard Allen, 2011: Grandfather E. C. Allen moved his operation across the street to the building now housing Attorney Parakilas' law offices and remained there until 1930 when the new store was built. Wilfred J. Starr submitted an estimate of costs to rebuild the store. The estimate was dated Oct. 18, 1930 was for $12,929.00. The store reopened on the original site and remained a community facility until 1993 when then owners Arthur G. Allen and his cousin, Edward C. Allen, II, principals and grandsons of E. C. Allen, I, dissolved the business amid growing competition and large corporations moving into Enfield. Allen's store closed its operation on September 11, 1993, after 111 years of serving the community. E. C. Allen, I, died in Dec. of 1933 at the age of 80. E. C. Allen, II, died on March 29, 2007 in his 80th year. At this writing Arthur G. Allen is the sole family survivor from the business.

The firm of

ALLEN & GOWDY,

is this day dissolved by mutual consent. The business will be carried on as heretofore by

E. C. ALLEN,

Mr. A. W. GOWDY can be found at the store as before, under the new firm.

Respectfully,

E. C. ALLEN.

Hazardville, Conn.,

July 1st, 1882.

Until further notice I shall close store every **THURSDAY EVENING** at 6 o'clock.

Very Respectfully,

E. C. Allen,

Hazardville, Ct. Oct. 10, 1882.

E. C. Allen's new store - circa 1930
Herbert W. Allen, standing at the west door

Scenes inside the 'new' store

Will Oliver – clerk – E. C. Allen's

Interior view of E. C. Allen's store – circa 1935

Twenty Mule Team Borax

Twenty Mule Team Borax, circa 1900, Main Street, Hazardville, Connecticut

This picture of a twenty-mule team was taken in front of the Methodist Church and in front of Edward C. Allen's home on Main St. in Hazardville, Conn. (circa 1900).

In a telephone conversation with a docent at the Twenty Mule Team Museum in California it was revealed that the twenty mule team made promotional tours around the country until about 1940. Since E. C. Allen's sold a large quantity of Borax, it can be determined that the mule team was in Hazardville as a result of Allen's sales.

Alice Godard Allen

The Charter Hotel

SYLVESTER F. CHARTER.

A prominent figure in the village of Hazardville was Sylvester F. Charter, who was born in Tolland, Conn. in 1825. When Sylvester was only 18, his father was drowned in a canoeing accident and the young man worked the next few years in the farming industry. Eventually he went to work in the powder mills and in 1851 he married Sophronia Page of Groton, Vermont.

In 1853 he began construction of the hotel and on March 4, 1854 Col. Hazard held a banquet at the new hotel for 150 employees. It is reported that the menu consisted of 15 turkeys, 6 roast pigs, 28 gallons of oysters along with other delicacies and champagne and it was at this event that the village of Hazardville was named for the president of the Hazard Powder Company.

The Prize Fight

The Charter Hotel

Almost lost in the history of the All American City (Enfield, CT) is the story of the secret prize fight at Charter's Hotel in Hazardville center (a/k/a The Hazardville Hotel). This clandestine affair was the brainstorm of a Springfield promoter, and was scheduled for Enfield because boxing was legislated against in that Massachusetts city. The names of the promoter and the participants have been forgotten, but word was passed around that the secret bout would be held in the third floor ballroom of the hostelry founded by Sylvester Charter at the corner of south Maple and Main Streets (now Hazard Ave.). It became so generally known, in fact, that the Springfield Police Dept. heard about it and telephoned the keeper of the general store nearby, to learn the date and location, as determined were they to raid the operation.

So it was that on a summer night in 1902 that E. C. Allen permitted his sons, Herbert and George, to stand in the shadows of the pillars on his front porch to watch the Springfield constabulary descent upon the illicit boxers. Herbert Allen's most vivid recollection of the night was recalled by him this week as being a severe shivering of his knees as the men in blue started up the stairs of the building now owned by Frank Skower. He recalls that some of the spectators got away and some were taken in by the police. The next morning the boys found a quantity of tiny black coal hode stashed in a fence corner at the opposite corner. These had been pinned on coat lapels of the spectators and craved as tokens of admission to the fight.

The Charter Hotel figured in other aspects of Enfield history since it was here that the proposal was made that the village in eastern Enfield should be named Hazardville in honor of Col. Augustus G. Hazard, prime mover behind the Hazard Powder Company. The local gunpowder concern is reputed to have produced as much as 40% of the munitions used by the Union during the Civil War. The old hotel is described as quite a community center by former Police Chief Earl Reynolds. The hotel office was in the front portion of the building now occupied by The Charles Luce Appliance Co. Behind the office was the barroom. One newspaper account details that a drummer, or traveling salesman, was knocked from a barstool in this room when an accidental explosion occurred in Hazard's nearby plant. The hotel dining room was located in the eastern part of the building and all lodging was on the ground floor with its long front porch.

Business activity of the hotel and of the village was punctuated by the hourly arrival of the trolleys of the Hartford and Springfield Street Railway Co., on which the young Earl Reynolds served as a conductor at the time of World War I. The ride from Thompsonville to Hazardville cost five, then six and finally seven cents, according to Reynolds. In addition to passengers, the line carried the mail, and freight as well. A switch was installed at the junction of Elm and Smoke Streets where east and west bound cars could pass. Trolley lines extended on to Scitico and Somers, and provided transportation for High School students from these areas. Another branch of the line ran from Warehouse Point to Ellington. Part of this right of way now is maintained as the East Windsor Trolley Museum.

The Melrose Trolley Trestle, circa 1906.

Home of Mr. & Mrs. Eli Stiles, Melrose, Connecticut.

Mrs. Stiles was the sister of E. C. Allen, I. "Lizzie" Stiles died of injuries sustained in a fall from the tower while washing windows and is buried in the Stiles plot in the Melrose Cemetery, diagonally across the street from this once imposing home, which stood on the corner of Broad Brook Road and Melrose Road. The garage in the rear still stands.

Trolley Accident on Main Street

These photos were taken by Herbert Walden Allen – date unknown. Location is Main Street, Hazardville, Connecticut, in front of the Methodist Church and the Allen residence.

Speech Given at the Hazardville Fire Department

Edward Channing Allen, II, Deputy Chief

Welcome to this celebration of 100 years of volunteer fire fighting in Hazardville. Hazardville, Conn., 100 years ago, was a community recognized throughout America. Memories of the Civil War burned brightly for every adult. Agents of the Hazard Powder Co. from Sutters Mill, California to San Antonio, Texas knew Hazardville as the source of the finest black gunpowder in the world. Even though Col. Augustus Hazard had died in 1868, peak production at the New Works on Cooper Street wasn't reached until the 80's. The powder industry had spurred rapid growth in Hazardville during the 1870's and 1880's.

One of Hazard's legacies was the idea of the Hazardville Institute – a social center providing a library, auditorium, clubroom and gym. This is in the area now recognized as a Federal Historic District.

By 1892 it was time to safeguard the infrastructure through a system of fire protection. In November 1892, William Henry Harrison was President of the United States. Another Civil War veteran, Andrew Gordon (Fred Gordon's great grandfather) chaired a meeting which voted to seek incorporation by the Legislature, and move forward to have 13 fire hydrants installed. Three commissioners were elected: Smith, Chapin and Allen. A three mill tax rate was set.

The first apparatus, a 12 man hand pumper proved inadequate, and was returned for a refund. It took four years of investigation before the coveted steamer could be brought in for testing. Hooked to the hydrant at the drug store, it successfully pumped water over the Methodist Church steeple, then over the spire of St. Bernard's Church.

Hazardville Fire Department's new steamer being tested at the Methodist Church.

Another view of the coveted steamer

The first fire station, built on this site in 1898 sheltered this $300 steamer. The first motorized engine, a Brookway, was added in 1919. When I joined the HFD in 1945 under Chief Raymond Smith and Dep. Chief Duffy Ryan, we worked out of this original 2 story building.

First station house of the Hazardville Fire Department

Fred Stroiney succeeded Chief Herbert Allen and a modern 4-stall station was built. Today it is our privilege to open this expanded facility and celebrate 100 years of volunteer service to the community of Hazardville.

The new Hazardville Fire Department
Picture by Chief Jack Flanagan
January, 2011

Fourth of July Celebration

According to folklore the Hazard Powder Company would make available a supply of gunpowder to Hazardville residents on the fourth of July. Residents would help themselves to the quantity they needed for their personal use and the men of the village would lay a line of gunpowder in the center of the road from Scitico to the center of Hazardville, about a mile.

Along the route, each person would have his cannon ready and lay his fuse from his front yard to the powder in the road until all was ready. From the center of Hazardville the signal would be given along the line – READY – READY – READY – until word reached the beginning of the line in Scitico. The line of powder would be lit and all along the route could be heard – BOOM!, BOOM!, BOOM! as the individual cannons were ignited, until the charge reached the center of Hazardville and the Tariffville Talker!! What a celebration it would be!

Note: The cannon known as the Tariffville Talker may have had other names. It was reported to have been submerged in one of the ponds of the Hazard Powder Company and has not been recovered or seen since.

The Hazard Powder Kegs

The Hazard Powder Kegs Baseball Team

Seated left to right: Bob Stimetz (P), Jerry Distazio (OF), Ray Knight (c), Richie Kwiecienski, (2B), Richie Wlodko (3B), *unidentified, unidentified,* Frank "Shankey" Arcesewski (SS), *unidentified,* Standing rear, *unidentified, And coach, unidentified*

The Hazard Juniors - 1950

Back Row (left to right) Steve Underwood, Arthur Felie, Bill Newport, Coach Arba Cooley, Arthur Bourque, Charles Lagel and Richard Taubner
Front Row (left to right): Rob Welch, Billy Hodgins, Dennis Balcewicz, Jake Congrove, Arba "Skip" Cooley, Arthur Newport, Russell Thrift, and Benny Debonis

Standing (rear) Manager Campbell and Coach Ryan
Middle Row (L-R) B. Berozsky, Girlie Davis, Helen Firtion, L. Tomalo, Ethel Firtion, D. Leger, M. White
Front Row (L-R) J. White, S. Skower, H. Miller, V. Pranka. Mascot L. Starr

Women's Basketball

1933

Back Row (L-R) Marion St. Louis, Ethel Firtion, Harriet Smith, Mary Savage, Helen Firtion, Coach Harold Firtion
Front Row (L-R) Helen Blonka, Lena Angelica, Babe Miller, Girlie Davis, Helen Wysocki

Back Row (left to right): Dick Copeland, Helen Firtion, Lena Angelica, Girlie Davis and Coach Charles Luce. Front Row (left to right): Ethel Firtion, Carolyn Locke, Babe Miller, and Harriet Smith

(L-R) Girlie Davis, Helen Blonka, Helen Firtion, Ruth Green, *Unknown, Unknown*, Ethel Firtion, Marion Jolly and Harriet Smith

The Enfield Inn

Colonel Hazard's home on Enfield Street, also known as Hazard Place, became the Enfield Inn, many years after the gunpowder industry was just a memory. In 1969, during a restoration, fire totally destroyed the once elegant home. Edward Allen was one of the Hazardville Volunteer Firefighters who battled the blaze.

**Hazard Hall (a.k.a the Enfield Inn) destroyed by fire.
Photo by Edward Channing Allen, II**

Menu from the Enfield Inn

ENFIELD INN

Enfield, Connecticut

1263 Enfield Street • Route 5

Telephone: RIverview 5-8850

Enfield Inn

Full Course Dinners

Chicken Cacciatore Served in Casserole	$2.25
Veal Scalopini With Mushrooms and Wine	1.75
Veal Cutlet Choice of Milanaise Style or Parmigiano	1.50
Broiled Sirloin Steak A large cut of prime heavy Black Angus	4.00
Broiled Filet Mignon A special extra thick cut of Tenderloin	3.50
Chopped Sirloin Steak Chopped and Grilled in Butter	2.00
Broiled Lamb Chops Two thick Lamb Chops Broiled to your Taste	2.50
Broiled Pork Chops with Apple Sauce Broiled Center Cut Jersey Pork Chops	1.75
Virginia Smoked Ham Steak Old fashioned goodness served with tangy Pineapple Ring	2.00
Roast Stuffed Turkey Cranberry Sauce and Turkey Gravy	2.00
Broiled Half Spring Chicken Broiled to perfection and served with a hot Butter Sauce	2.00
Southern Fried Spring Chicken Dipped in Egg Batter and Fried to a Crisp Golden Brown	2.00

THANK YOU ::

A La Carte

Fifty Cents Extra

SEAFOOD

Shrimp Cocktail	1.00
Broiled Sword Fish Broiled to tenderness served with chopped Parsley and Butter	1.50
Broiled Halibut A thick Steak, Lemon and Butter	1.50
Broiled Lobster Broiled Live Lobster served with drawn Butter	3.00
Lobster a la Newburg Served in Casserole with Toast	2.50
Lobster Salad Fresh Lobster meat, chopped Celery and Mayonnaise	2.50
Fried Shrimp Tartar Sauce and Lemon Wedge	2.25
Shrimp a la Newburg Served in Casserole with Toast	2.25
Shrimp a la Creole Served in hot Casserole	2.25
Frogs Legs Saute Meuniere or Provenciale	3.00
Fried Scallops Deep Sea Scallops dipped in a rich batter and fried to a golden brown	2.00
Fried Filet Haddock Served with Tartar Sauce and Lemon	1.75

CALL AGAIN

Hazardville Institute

The Hazardville Institute was the brainchild of Col. Augustus G. Hazard of Enfield, founder of the black gunpowder plant, which bore his name. In the 1820's Hazard was a dealer in paints, oils and sundries with packet ships trading between New York and Savannah, Georgia. He soon became an agent for the gunpowder being produced in Enfield by Allen and Parkes Loomis of Suffield. Hazard suffered reverses during the financial crash of 1837, so from his base in New York decided to move into the powder business. Reestablishing his credit, he moved to Enfield and bought a part interest in the Loomis brothers concern. Shortly thereafter Hazard became sole owner of the powder company.

Col. Hazard took a rather paternal attitude with his employees and their village. For example, since many of his experienced powder makers had been brought over from Faversham, England, Hazard, a Congregationalist, felt a responsibility to build an Episcopal church in the village which was served by only Methodist and Catholic churches. "Col. Augustus Hazard, president of the Hazard Powder Company, gave the church and the land on which it stands, and an endowment of $5000 for its maintenance" – Parish History

As the war ended, Hazard felt the need for a social and recreational facility to be available for the whole community. St. Mary's, dedicated September 29, 1864, had no parish house. Neither did Hazardville Methodist Church or St. Bernard's Roman Catholic Church. The Hazardville Institute was to embody his desire for a social and cultural center.

A "Charter of Hazardville Institute, Sept. 1868" was passed by the May session of the Connecticut General Assembly. These papers of incorporation resolved that Hazard and others, mostly his employees, be constituted "....for the purpose of erecting a building in the village of Hazardville, town of Enfield, to be devoted to moral and intellectual improvement and for promoting other enterprises having the same intent and purpose and by that name shall have perpetual succession...."

The list of subscribers is headed by Hazard who took 200 shares (par value $10) and five of his closest associates whose shares totaled $3000. Hazard also gave the lot of land on which the brick building was constructed. The corporation officers included a Board of Directors, but could not include Col. Hazard who had died at the Astor House (1868) in New York City shortly before completion of the structure in 1869.

The records of the treasurer do not disclose the Institute to be self supporting, and the stockholders were frequently assessed to keep the building open. Over the years, nominal rental income shows varied community use. A village library preceded the town wide public library system. The above mentioned churches used the hall frequently, as did the Enfield Grange, Hazardville Brass Band, Enfield school system for extra classrooms, Men's basketball teams, public polling place, Volunteer Fire Company annual banquets, school graduations, Red Cross classes during World War I, etc. In the period 1915 – 1920 a series of public lectures were held. Among those speaking were Russell Conwell with his "Acres of Diamonds" oration. One attendant remembers there were 306 people in the audience. Variety shows and little theater flourished.

After the 1940's with the expansion in population and construction of social and classroom facilities by the churches, the Institute saw less use. On Dec. 14, 1956, the stockholders voted to sell the property to the Town of Enfield for $1000 with the stipulation that it be used for a Youth Center. The two downstairs rooms housed woodworking and craft activities, and a basketball court filled the upper assembly hall. Central heating was installed to replace the two large coal stoves and registers, and rest rooms replaced the earlier outdoor facilities.When Hazardville was registered as a Federal Historic District, Hazardville Institute was placed on the list of registered historic buildings. With the creation of the nearby Scantic River State Park in Powder Hollow, the Institute is ideally suited as a volunteer staffed headquarters and trail base for the park. Augustus Hazard would be pleased.

OF THE

Hazardville Fire Company

AT INSTITUTE HALL

WEDNESDAY EVENING, FEBRUARY 22, 1911

At 7:30 O'clock

JOHN SHEERN

TOASTMASTER

TOAST

PLEASANT MEMORIES

Let Fate do her worst, there are relics of joy,
Bright dreams of the past, which she cannot destroy.
They come in the night-time of sorrow and care,
And bring back the features that joy used to wear.

Long, long, be my heart with these memories filled,
Like the vase in which roses have once been distilled;
You may break, you may shatter the vase if you will,
But the scent of the roses will hang round it still.

Thomas Moore

MENU

Escalloped Oysters

Roast Turkey *Mashed Potato*

French Rolls

Queen Olives *Little Midgets*

Neapolitan Ice Cream
Assorted Cakes

Coffee *Cigars*

Page Two

"Lest we forget."

Auld Lang Syne

Should auld acquaintance be forgot,
  And never brought to mind?
Should auld acquaintance be forgot,
  And days of auld lang syne?
 CHORUS.–
For auld lang syne, my dear,
  For auld lang syne;
We'll tak' a cup o' kindness yet
  For auld lang syne.

We twa ha'e run aboot the braes,
  And pu'd the gowans fine;
But we've wandered mony a weary foot
  Sin' auld lang syne.
 CHO.–

We twa ha'e paid'let in the burn,
  Frae mornin' sun till dine;
But seas between us braid ha'e roared
  Sin' auld lang syne.
 CHO.–

Page Three

Officers

ALBERT F. BAKER, President.
ARTHUR G. GORDON, Vice President.
WALTER E, ALBEE, Secretary.
WILLIAM SKINNER, Treasurer.
GEORGE J. GORDON, Foreman.

Committee of Arrangements

William W. Gordon,
Charles Abbe,
David A. Bridge,
William Morris,
Edward Locke,
George H. Payne.
William A.Bridge,
William Skinner.

Back Page

Hazardville Institute Conservancy

The Hazardville Institute Conservancy, Inc. involves Hazardville folks dedicated to the historic Hazardville Institute as a local social, historic and cultural center. In the early 1980's the Conservancy raised $20,000 and succeeded in rebuilding the roof structure and chimneys, Now it is time to complete the renovation.

The Hazardville Institute, a vision of Col. Augustus G. Hazard, was created as a community center in 1868. It served this purpose for 100 years before the stockholders turned it over to the town with the stipulation it become a youth center.

Neglected for years, it was threatened with demolition, a sad end for a landmark listed on the Federal List of Historic Places, the focus of our village Historic District.

The Conservancy's goal is to reopen the building as a volunteer headquarters and trail head for the Scantic River State Park, on the nearby site of Col. Hazard's Powder Company. It would include a museum to that industry, with exhibits, as well as meeting space for community organizations.

Today the Conservancy's committee is working tirelessly to restore the building. The Hazardville Institute Conservancy Society's Mission Statement is "To preserve and sustain the Hazardville Institute, a landmark building integral to the Village of Hazardville and the Town of Enfield, as a civic meeting place which will foster, encourage and promote cultural, historical and intellectual enlightenment and activities."

Through generous corporate and individual donations and the ongoing work of a dedicated committee that our goal will be realized.

And the work begins

Gilbert Vasseur, left, & Edward C. Allen, II surveying the work at the institute.

Ed Allen and Gretchen Pfeifer cleaning panes of glass.

There have been numerous fund-raisers and manual labor including collecting old windows and glass from demolished homes, cleaning the glass and cataloging and storing it by age and quality. Enough glass was collected to restore all 23 windows at no cost and have plenty left over.

Windows before and after restoration

Stenciling was discovered in 2006 just before the plaster walls were to be removed so work was delayed to allow time to scrape and reveal the stencils, trace the patterns and cut out samples for future study. Fundraising letters were sent in 2006 promoting window commemoration sponsorships and raised over $30,000 to date.

Commemorative fund raising items

Hazardville Clock

One of a series of ornaments

Ed Allen and Bob Tanguay advertising the Halloween Ghost Walk held at the Hazardville Cemetery. Ed Allen appeared as Col. Nathan Axtell at the annual ghost walk

A bus tour to the Hancock Shaker Museum in Hancock, Mass. was held as another fund raiser for the Hazardville Institute. Gretchen Pfeifer, President of the Hazardville Institute Conservancy is on the left, with Alice and Ed Allen, tour directors. Subsequent tours to the Hagley Museum in Wilmington, Del. and to the battlefields of Antietam and Harper's Ferry were offered to the public to benefit the institute. Karen LaPlante, not pictured, was also involved in the planning.

In 2007, following the death of Edward C. Allen, II, one of the original committee members, memorial donations given in Ed's memory were earmarked for the restoration of the front and rear doors. A public dedication ceremony was held on May 23, 2010.

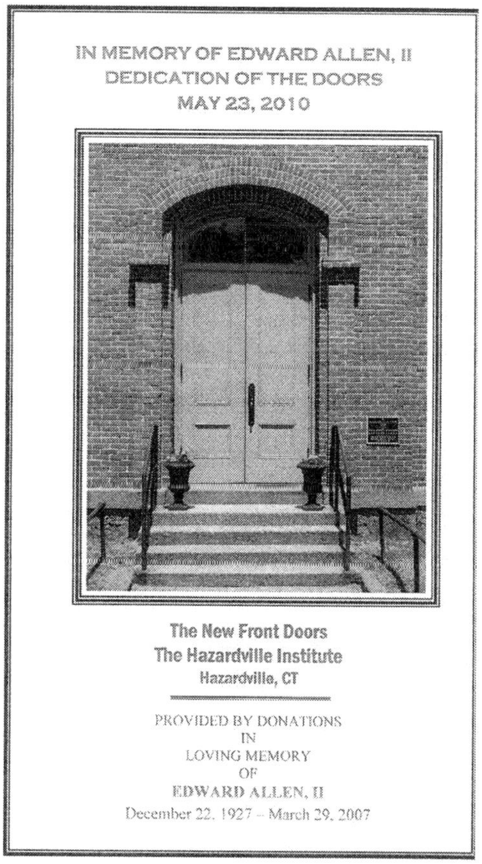
IN MEMORY OF EDWARD ALLEN, II
DEDICATION OF THE DOORS
MAY 23, 2010

The New Front Doors
The Hazardville Institute
Hazardville, CT

PROVIDED BY DONATIONS
IN
LOVING MEMORY
OF
EDWARD ALLEN, II
December 22, 1927 – March 29, 2007

The new front and rear doors of the Hazardville Institute

Hazardville Grammar School – then and now

View of Main Street, Hazardville, Connecticut looking west

View of Main Street, Hazardville, Connecticut looking east

Fairlawn Avenue (now School Street) looking North. The house on the right is located approximately where Oak Street is today. That house was later moved and is today 5 & 7 Oak Street. School Street was then extended to intersect with what is now North Street.

Fairlawn Avenue (School Street)
Hazardville, Connecticut looking north

Fairlawn in Winter

Entrance to the Manager's house – driveway west of St. Mary's Church

Two scenes on Main Street, Hazardville, CT – pre 1919 – looking west

Early photo of Bridge's office and factory, Hazardville, Connecticut

Gordon Brothers Mill, Scitico, Connecticut

The following pictures were taken from glass negatives – most taken by William W. Gordon in the late 19th century and/or early 20th century. Not all are identified but we include them for their individuality. The names are listed where they are known.

This picture is taken in front of 23 School Street, home of Dr. John Bridge. The men are not identified but one might assume that Dr. Bridge is one of them – perhaps the gentlemen in the middle.

Emily Bridge, left, and Miss Killam in front of the schoolhouse at Scitico, CT

Amos Down Bridge and his wife, Elizabeth Gordon Bridge

Unidentified occupants of the house on School Street on the northeast corner of Oak Street (circa 1900)

Parades

Parade heading west on Main Street, Hazardville, Connecticut

Parade heading west on Main Street, Hazardville, Connecticut Building on right is the first Methodist Church built in 1835.

Boys in race – heading west on Main Street

Pictures by George E. Allen

Parade heading east in front of the Hazardville Fire Department

Parade in front of Hazardville Fire Department – heading east

Parade heading east just past the Hazardville Fire Department

Hurricane Carol – 1955

In the aftermath Hurricane Carol, 1955, Edward received a phone call saying that the Scantic River at South Maple Street was rising to the level of the bridge. Ed grabbed his camera and along with his cousin, Arthur Allen, rushed to the scene in time to capture these two images.

Arthur Allen watching the Scantic River raging against the bridge at South Maple Street after Hurricane Carol, 1955.

South Maple Steeet Bridge in the aftermath of the hurricane of 1955.

The following pictures of Hurricane Carol, 1955 were taken by David Niel Jellison.

South Maple Street Bridge looking south – Hurricane Carol, 1955

South Maple Street Bridge looking south

South Maple Street Bridge looking north

South Maple Street Bridge from the sand cliffs

Destruction of Hurricane Carol near the site of the New Works

Powder Hollow – Back to Nature
Photos by Edward C. Allen, II

Along the Scantic River

Scantic River Spring Splash

Scantic River Spring Splash

Scantic River Spring Splash

Grosbeak

Chickadee

The Scantic River in Morning Light

Remains of the Upper Falls

The Scantic River at Dawn

South Maple Street Bridge

Hazardville Church Histories

St. Mary's Episcopal Church

St. Mary's Memorial Plaque to honor Col. Augustus G. Hazard

Excerpts from the Parish History

The cornerstone for St. Mary's Episcopal Church was laid on May 11, 1863, by the Rt. Rev. John Williams, then Assistant Bishop of Connecticut. In it were placed the Holy Bible, the Book of Common Prayer, Diocesan Journal, Minutes of the meeting for organization, list of parish officers, Church Calendar, Hartford Times, and the Springfield Republican.

The building was consecrated by Bishop Williams of the Feast of Saint Michael and All Angels, September 29, 1864. The whole cost of the church and appurtenances, including the bell, upholstery, the sheds and fences, was about $13,000.

On March 25, 1864, the Rec. C. Collard Adams became Rector and remained for a little over a year. In 1871 the Rev. L. Simonson took charge and from 1872 until 1898, several priests came and departed. They included the Rev. David P. Sanford (1874-1882), the Rev. J. B. Robinson (1883-1892), the Rev. J. Francis George (1885-1892), the Rev. T. D. Martin (1892-1897).

The Rev. D. Russ Judd, well remembered by many of our communicants, became Rector in 1898 and remained until February 1, 1920.

On Sunday, October 4, 1914, the fiftieth Anniversary of Saint Mary's Church was observed. A memorial brochure was prepared and distributed.

From 1920 until 1925, the Rev. Glenn B. Coykendall, the Rev. E. T. Mathison, and the Rev. John Wilkins were successively in charge. The Rev. William P. Downes assumed charge in 1925 and remained until 1937 when he was succeeded by the Rev. Francis R. Belden. He served until 1947 when the Rev. John Hall Findly became priest-in-charge. Father Findly remained until 1951 to be succeeded by the Rev. Roy MacKaye Atwood, who served until 1957.

Saint Mary's Church came into being in the heyday of the Hazard Powder Company. The number of Episcopalians residing in the community necessitated the establishment of Saint Mary's Church. With the departure of the Powder Company from Hazardville, and the subsequent decrease in population (c. 1914) the Parish was considerably weakened both in communicant strength and in financial resources. Had it not been for the loyalty and devotion of the small band of Episcopalians who were determined to keep their church open, and the part-time assistance of the priests from Saint Andrews, Thompsonville, Saint Mary's would perhaps have passed quietly from the scene. Not so, however, and the round of Prayer Book services was maintained until the "building boom" of the mid 50's brought to the community a tremendous population increase. The Episcopal Church was here, to continue the mission for which is was established a hundred years ago.

As the Parish began to grow once again, it became increasingly evident that a resident priest would be necessary. Consequently, in 1958 the house and property immediately to the rear of the church was purchased for a rectory, and on September 14, 1958, the Rev. Robert W. Watson was instituted as Priest and Rector. Since 1957 an annex has been added to the church, a Parish House was built in 1961 and the old rectory was replaced by a new one, built on the same site.

All present indications are for a continuing population and building growth in the community for the next several decades. Saint Mary's Parish has a ministry to fulfill, and by the grace that our Blessed Lord gives us, we shall continue to "work and pray and give for the spread of His Kingdom," not only in our immediate community, important and necessary as it is, but throughout all of God's world.

St. Bernard's Roman Catholic Church

Parish History

Until 1852 when Hazardville passed to the care of Windsor Locks, the section was tended by Hartford priests. About 1860, the Maple Street home of William Casey became the site of Hazardville's first Mass. The celebrant was Father James Smyth of St. Mary, Windsor Locks. In January 1863, Hazardville became the pastoral responsibility of St. Patrick parish, Thompsonville. In 1865, with an old schoolhouse outfitted for Mass, formal worship came closer to reality in Hazardville. Father John A. Mulcahy proceeded to build a church on Main Street. Bishop Lawrence S. McMahon blessed the cornerstone on September, 12, 1880, and the church was dedicated on July 17, 1891. In 1888, Father John Synnott became the first resident pastor of St. Bernard as it was raised to parish status by Bishop McMahon. Father Edward J. Reardon erected a new church on the site of the demolished old one, now too small to accommodate the congregation. Archbishop Henry J. O'Brien officiated at the dedication in 1956. Father Reardon also erected a parish school in 1958 on Hazard Avenue, staffed by the Felician Sisters.

Hazardville, Conn.

St. Bernard's Roman Catholic Church and Rectory.

ALLEN & SKINNER, PUBLISHERS 50C4.

The Hazardville Methodist Church

An early history of Methodism in Enfield was written by the Reverend Charles W. Gallagher who was Pastor from 1884 to 1886. Early Methodist meetings were held on Enfield St. in a building located at the corner of what is Oliver Road. In 1829 a camp meeting was held on Job's Hill in Somers which eventually resulted in meetings being held on the east corner of the present Holiday Lane and Hazard Avenue. In 1835 the first church was built in Hazardville – the building, which is now an apartment building, still stands opposite the 'old' Hazardville cemetery on Hazard Avenue.

Hazardville Methodist Church – built in 1835

In 1872 the present brick church building was erected at a cost of $23,000, of which $18,800 had been subscribed by the people and $2,600 was realized from the sale of the Old Church property. In 1885 the debt was paid off, the church was redecorated and new carpet was purchased. The choir loft was directly in front of the large stained glass window we see on the front wall of the church. It was a raised, semi-circular platform with a railing around it with chairs for the choir. The organ was a cabinet organ used until 1885 when a pipe organ was bought for $1,000.

Original cabinet organ and choir loft on the north wall of the sanctuary (c. 1909). In 1909 the church was wired for electric lights, replacing a system of gasoline lights. Kerosene lamps along the side walls were used.

Hazardville Methodist Church Sanctuary after installing electric lights

In 1923, a three story addition was added on the south side of the church that provided space for a Ladies parlor, classrooms, a kitchen and a heating plant.In 1924, Mrs. George B. Gordon gave a new Estey Pipe Organ and Mrs.David Gordon gave the pictorial stained glass windows, The installation of the organ necessitated the rearranging of the sanctuary and moving the choir to the altar area in front of the congregation. It also provided an organ chamber for the pipes in the southwest corner of the sanctuary and a corresponding preparation room in the southeast corner to balance the room.

Further decoration included adding imitation pillars and arches to the altar area with the communion rail joining the pillars. Also, the balcony was erected during the 1924 renovation.

About 1945 the Gilbert Pierce room was excavated. Gilbert was the first war casualty of the parish, killed in Belgium in December 1944, at the age of nineteen. The excavation was done with volunteer help from the parishioners.

The next large improvement was the building of Fellowship Hall and the Sunday School complex in 1961. This meant the removal of the parsonage which was located next door to the church, on ground now occupied by Fellowship Hall. The church bought a lovely home on Elm Street belonging to Clarence Martin to be used as the new parsonage. It provided space large enough for a family and was within walking distance of the church property.The new addition provided Sunday School rooms, a large social hall, offices for the pastor and church secretary as well as a new kitchen and activity rooms.

In 1972, during the 100th anniversary of the building of our church, a committee was established to study sanctuary renovation. The committee met many times but had no real success or direction. Many ideas were formed but nothing was ever accomplished and no work took place. However, a fund drive was started and $20,000 was raised for renovation.

It wasn't until 1984 that the church embarked on a major renovation of the sanctuary, the purpose of which was to enlarge the chancel area and give greater space to the worship center and choir loft. The new committee decided that a reasonable expenditure would be the hiring of a church architect so we authorized $4,000 to be spent out of our renovation fund. There was a mural of Jesus seated at the Sea of Galilee on the back wall of the church, in the chancel area, a donation from our longtime organist Mae Pierce. The picture became a casualty of the renovation, much to the dismay of about 50% of the congregation. However, Mae Pierce was heard to say, many years prior, that if the church ever wanted to take the picture down it was okay with her. We compromised by having a picture of the mural taken and copies made available to anyone interested in having one.

The reason for removing the mural was to move the pipes for the pipe organ from the west wall of the choir loft to the south wall, directly behind the choir and in full view of the congregation. This was necessary in order to open the wall of the organ chamber to allow more sound to come through when the organ was being played.

The entire sanctuary was newly carpeted as a result of the renovation. The old communion rail could accommodate only 15 people at one time and the new rail would seat more than 30. The choir loft was also enlarged to seat 27 people and the organ was moved forward a few feet. Two sections of the old communion rail were used for the new rail and the new railing was built to match the old. Also, when the pillars and arches were removed another ceiling beam, which had been completely covered by the 1924 renovation, was revealed. The cornices were missing and the carpenter made a template to match the existing cornices. There is another beam still covered in the ceiling above the choir.

During the three months of the renovation of 1984, Sunday morning church services, including Easter of that year, were held in Fellowship Hall. One wedding was held in Fellowship Hall during the renovation.

In 1994 the Estey Pipe organ was failing and the church decided to replace it. We contacted seven different companies to submit proposals to the committee.

After narrowing the selection down to three we, the committee, made a proposal to the Administrative Board for the purchase of an Ahlborn-Galante combination electronic/pipe organ at a cost of $52,000. The Ad Bd voted in favor and the purchase was made.

The sound of the Ahlborn-Galante organ is accomplished by what is called digital sounding. They take a sounding of one particular pipe from a pipe organ that could be in Germany and reproduce the sound digitally to produce the exact replica on the electronic organ. That technology is used to produce all of the organ sounds, with amazing results. This new technology allowed the Ahlborn-Galante company to connect 6 ranks of our existing pipe organ to the new electronic technology to produce a combination of sounds. We can play the electronics, we can play the pipes or we can play them in combination.

As we look to the future, there will undoubtedly be other changes taking place in this special house of worship. Only a vital and active community of people keeps growing and changing.

Alice Godard Allen, Historian
Hazardville United Methodist Church
Hazardville, Connecticut

Acknowledgments

My thanks first to our children, **Edward Channing Allen, III**, **Laura Rose Allen Fiore** and **Dr. Beth Janette Allen Moore**, who have given my life purpose, who have shown love, respect and admiration for their father and me and who have made our lives worth living. They will never know the full scope of the joy their presence in our lives has meant to us and how they have enriched and enlivened us. Their constant love and support gave me the courage to finish this book.

Ted Bayck – for allowing us to use photos of powder cans from his book "Gun Powder Cans and Kegs."

Milton Brown, PhD – for reviewing the manuscript.

Robert "Bob" Clark – for his help and computer knowledge.

Jeff Dynia – who spent countless hours scanning glass negatives and slides and other pictures for this book.

Mark Gordon – Author of his family genealogy and for his friendship and encouragement.

Hagley Museum and Library – For their assistance in my research regarding Col. Hazard and the Hazard Powder Co.

David "Whitey" Jellison – a longtime friend, for his support and contributions including permission to use portions of his college thesis on the Hazard Powder Company.

Karen LaPlante – for trips to Wilmington to research the Dupont files for information necessary to complete this story and for her friendship.

David McMahon author of 'Peril in the Powder Mills' for his ongoing help friendship, encouragement and advice. His knowledge and experience as an author was invaluable to me for this project.

Marjorie McNinch – Reference Archivist, Manuscripts and Archives Dept., Hagley Museum and Library for her assistance in researching material for this book.

Russell Meyer – For providing pictures of the Enfield Congregational Church whale oil lamp and altar columns.

Marjorie Bridge Nadeau – a very special thank you, for keeping the pressure on to finish this book. My undying gratitude for her tenacious and resolute determination in pursuing the completion of this venture.

The Rev. Craig Peel – former pastor of Calvary Presbyterian Church in Enfield CT, for his summation at the close of the Hazard Powder Company portion of this book .

Tom Phelan – For his assistance and computer skills.

Jean McGuire Smith – for helping to name the women's sports teams.

Richard Kweiceinski, Donald Sauer &Richard Taubner – classmates of Alice Allen, for helping to identify the Hazard Powder Kegs and Hazard Jrs.

Made in the USA
Monee, IL
13 September 2026

62977543R10168